Energy-saving Principles and Technologies for Induction Motors

# Energy-saving Principles and Technologies for Induction Motors

*Wenzhong Ma*
*China University of Petroleum*
*Qingdao, China*

*Lianping Bai*
*Beijing Information Science & Technology University*
*Beijing, China*

WILEY

机械工业出版社
CHINA MACHINE PRESS

This edition first published 2018 by John Wiley & Sons Singapore Pte. Ltd under exclusive licence granted by China Machine Press for all media and languages (excluding simplified and traditional Chinese) throughout the world (excluding Mainland China), and with non-exclusive license for electronic versions in Mainland China.



*Registered Offices*
John Wiley & Sons, Inc., 111 River Street, Hoboken, NJ 07030, USA
John Wiley & Sons Singapore Pte. Ltd, 1 Fusionopolis Walk, #07-01 Solaris South Tower, Singapore 138628

*Editorial Office*
1 Fusionopolis Walk, #07-01 Solaris South Tower, Singapore 138628

For details of our global editorial offices, customer services, and more information about Wiley products visit us at www.wiley.com.

Wiley also publishes its books in a variety of electronic formats and by print-on-demand. Some content that appears in standard print versions of this book may not be available in other formats.



*Library of Congress Cataloging-in-Publication data applied for*

Hardback: 9781118981030

Cover design by Wiley
Cover image: Courtesy of Wenzhong Ma and Lianping Bai

Set in 10/12pt Warnock by SPi Global, Pondicherry, India

Printed in Singapore by C.O.S. Printers Pte Ltd

10 9 8 7 6 5 4 3 2 1

# Contents

## About the Authors

**Wenzhong Ma** received B.S. and M.S.E. in Electrical Engineering from the Harbin Institute of Technology (1995), and Ph.D. in Electrical Engineering from the Institute of Electrical Engineering, Chinese Academy of Sciences (2006). Since 1995, he has been a faculty member at the China University of Petroleum, where he is a professor in the Department of Electrical Engineering.

He has taught a wide range of courses about electric machinery and power electronics, including Electric Machinery and Drive Systems, AC Variable Speed System, Power Electronics, Electric Circuit Analysis, and Electrical Energy-saving Systems.

He has done extensive research in electric machine and drives, including motor design, motor drives, converters, electric power-saving systems, and power electronic systems. From 2002 to 2006, he was involved in a national key project for Shanghai high-speed maglev train systems, which is the first commercial high-speed maglev train. He took charge of the commissioning and testing work of the long stator line motor, propulsion system, and power distribution systems. He has fulfilled the optimization of the long stator line motor system which is part of the Key Projects of the National High Technology Research and Development Program of China (863 Program).

He has authored five books: *Energy Saving Principle and Technologies of Electric Machinery* (China Machine Press, 2012), *Electric Machinery and Drive Systems* (China University of Petroleum Press, 2009), *AC Variable Speed System* (China University of Petroleum Press, 2013), *Analysis of Advanced Electric Circuit* (China University of Petroleum Press, 2010), and *Experiment and Learning Guide of Electric Circuit* (China University of Petroleum Press, 2007). He has published more than 40 papers, and he is also the inventor of three Chinese patents.

**Lianping Bai** graduated from Fuxin Mining College in 1982. He received M.S.E. in Electrical Engineering from the Harbin Institute of Technology (1990) and Ph.D. in Electrical Engineering from Xi'an Jiaotong University (2000). He was promoted to professor in 2001.

From 1982 to 1987, he served as a lecturer in Heilongjiang Mining college, where he taught courses on electrical automation in mining.

From 1990 to 1997, and from 2000 to 2005, he served in the China University of Petroleum; he taught a wide range of courses in Electrical Engineering, including Circuit Analysis, Electric Machinery and Drive Systems, Automatic Control System of the Electric Drives, Motor Energy-saving Technology in Oil Field for undergraduates, Computer Control Technology of Electrical Drives, and Principle and Application of DSP for graduate students.

He has been working at Beijing Information Science and Technology University since 2006. He teaches Electrical Engineering courses, including Circuit Analysis, Motors and Drive, Motor Energy-saving Technology, and so forth, for undergraduates; and Computer Control Technology of Electrical Drives, Motor Energy-saving, and Testing Technology for graduate students.

From 1993, Mr. Bai engaged in the research on the principle of motor energy saving in oil fields and published 13 articles in this aspect. He has finished several research projects such as the research on double-power energy-saving motor and its control device, the research on energy-saving technology with pole-changing motors for water injection pump in an oil field, the research on winding-type energy-saving motor and its control device, and the development of on-site testing technology and software for motors with a beam pumping unit.

He holds two Chinese patents: Beam Pumping Unit Energy-saving Motor and Double-Winding Series Energy-saving Motor. These two patents (CN96249172.1 and CN99220167.5) are now widely applied in the Shengli Oil Field and other oil fields in China.

# Preface

China has become a country of greater energy consumption nowadays, and energy supply shortage is getting worse and worse. This situation affects the economic development of the nation. Whatever the shortage is, there are still a lot of low-usage and waste of energy. Therefore, energy saving is a long-term policy for our social and economic development, as well as an urgent task for now. The government has set motor energy saving as one of the most important projects in the State Development Planning. This project requires technology and qualified scientists and technicians to implement the projects. Cultivating talents and training qualified personnel should start from university education. So it is necessary to set up a Motor Energy-saving Course in the Department of Electrical Engineering and Automation. Nevertheless, there is no relevant textbook on this subject, even though a few relevant reference books can be found now. Therefore, the author has written this book based on the handouts of Motor Energy-saving Technology and more than 10 years of study on motor energy saving. It can also be used as a reference book for workers engaged in this area.

The proportion of the installed capacity of a three-phase induction motor makes up more than 80% of the total installed capacity of the motor. Consequently, the book mainly discusses the energy-saving principle and method for three-phase induction motors. At the same time, as the permanent magnet motor is an important developing direction of efficient motors, the book also introduces the principle and the application of efficient permanent magnet synchronous motors.

People often mention the power waste of motors, such as "big horse pulling small cart." How can we define these problems? The book studies the variation rule of the motor efficiency and comprehensive efficiency curves, and defines the boundaries of the "big horse pulling small cart." It also analyzes the variation of induction motor losses in the case of the "big horse pulling small cart." The book also proposes some new ideas and new methods for induction motor energy saving. For example, the energy-saving principle of a double power motor, and combined pole-changing control for motors, and soft characteristics match. The field motor loss test method and the performance evaluation method for operating motors are also discussed. The efficiency reference value of rejection for used motors is also given. It puts forward the method of comparison evaluating for motor energy-saving effect. It provides the curves of active power and reactive power and power factor for beam pumping unit motor. It puts forward the experimental research methods of motor reactive power compensation, as well as the method of motor hybrid dynamic reactive power compensation.

The book combines the motor energy-saving principles, methods, techniques, and experience together, which shows the skill and experience to the reader, in order to enable readers to apply what they have learned. The book expounds the motor energy-saving principle through the power and load match, speed match, and mechanical property match. First, the book introduced the three-phase induction motor works and energy-saving principle, then the motor energy-saving methods with examples, and finally motor testing methods and evaluation methods. National standards are consistently reviewed in the book, in order to enable the reader to grasp the principle of motor energy saving, and in the meantime to be able to understand the standards and the usage of standard basis to follow when the instance and program options.

The motor energy-saving technique is developed not only to increase the efficiency of the motor itself but also to improve the efficiency of the motor drive system. The motor energy saving is a broad field; the motor drive system involves many aspects. It can be noted that motor energy saving is a complicated system engineering. Improvements have been made in motor energy saving over the past decade, but it still has a long way to go. For instance, research on motor design theory; improving motor manufacturing processes; development of motor manufacturing materials; research on motor control technology; developing the technology of motor drive; and research on motor drive system theory.

Chapters 1, 3, and 4, and Sections 5.1, 5.5, and 5.6 are written by Lianping Bai, from Beijing Information Science & Technology University. Chapters 2, 6, and 7, and Sections 5.2–5.4 are written by Wenzhong Ma, from China University of Petroleum.

We would like to sincerely thank Professor Yanmin Su of Xi'an Jiaotong University for his review and valuable opinion during the preparation of the book. We would also like to thank the authors of references cited in the book. We would like to express our appreciation to Liang Zhang, Chun Zhang, and Yongliang Liang for their contribution to the book.

There may be errors or improper expressions in the book due to the limited knowledge of the authors. We sincerely welcome feedbacks and suggestions, which can be sent to mawenzhong@126.com. Thanks!

*Wenzhong Ma and Lianping Bai*
June 2016

# About the Book

This book focuses on energy-saving principles and methods for three-phase induction motors, including the principle of energy saving in motor power and load match, speed match, mechanical property match; and the energy-saving methods for dual power motors, variable-frequency drives, combined pole-changing control, and soft characteristic match. The book also introduces the methods of motor reactive power compensation, the application of efficient permanent magnet synchronous motor, the economic operation analysis and energy-saving calculation of a three-phase induction motor, and the field testing and evaluating methods for motors. In addition, the economic operation and the energy efficiency standards of the three-phase induction motors, as well as a brief introduction of motor test platform, instrument, and procedure, are also covered in the book.

Through theoretical analysis and case studies, the book provides the energy-saving principle and techniques for induction motors. It can be used as a textbook for undergraduate or graduate students majoring in electrical engineering and automation, and it can also be considered as a reference book for electrical engineers.

# 1

# Introduction

## 1.1 The Energy-saving Status of an Electric Motor System

### 1.1.1 Basic Situation of an Electric Motor System in China

China is a great developing country with 20.43% of the total world population, but its energy per capita is lower than half the world's average level. China has become a high energy consumption country in recent years and the economic development is constrained to the tight energy resources. To make matters worse, the problems of low energy efficiency and extravagant waste of energy still exist in the country, which cause serious environmental disasters. In China, energy efficiency is about 10% lower and energy consumption per unit of energy-intensive products is 45% higher than world lead level, respectively. Data have shown that energy consumption per $10,000 GDP of China is a lot higher than that of Japan, Germany, or the United States.

In the world today, the transfer from electric energy to mechanical energy is basically through electric motors, and hence the electric motor supplies the most mechanical energy. By the end of 2010, the total installed capacity for various kinds of electric motors in China was 650 million kW and they occupied half the total power generation. However, the average rated efficiency and average operation efficiency of electric motors in China are 3 and 5% lower than those of developed countries, respectively, which cause a waste of electric energy of about 100 billion kWh per year with an average annual operation time of 3000 h. All the aforementioned data show that the energy-saving technology of electric motors badly needs development in China.

The development of energy-saving technology of electric motors in China faces two main problems.

One problem is the low efficiency of electric motors. The development of manufacturing technology for electric motors is greatly constrained to vicious market competition. Many manufacturers used to take low cost as their primary consideration instead of high efficiency. For example, a permanent magnet motor has 5% more efficiency, but the high cost restricts its market share. In recent years, even though energy efficient motors and rare-earth permanent magnet motors have been greatly promoted, the development is still slow.

The other problem is that the power match between motors and loads is not appropriate. Most electric motors are in low-efficiency operating conditions because of the mismatches of mechanical equipment and electric motors, including unreasonable lectotype, capacity, torque, and rotation speed. For example, large power

*Energy-saving Principles and Technologies for Induction Motors*, First Edition. Wenzhong Ma and Lianping Bai.

motors, like water pumps and compressors, have an efficiency lower than 50%. Besides, design margin exists in every step for a drag system, which further decreases the efficiency.

### 1.1.2 The Main Contents of Energy Saving for Electric Motors in China

Ten key energy-saving projects were started in 2006 by the National Development and Reform Commission of China. As one of the key projects, energy saving for the electric motor system covered the following aspects:

1) Upgrading or Eliminating Low-Efficiency Electric Motors and High Power Consumption Equipment
   Promote high-efficiency electric motors. Limit and forbid the manufacturing, sales, and utilization of low-efficiency products gradually. Upgrade the old equipments, including reforming high power consumption medium and small motors and pumping system, reasonably matching the constant flow hydraulic system.
2) Improving the Electric Motor System Efficiency
   Promote innovative speed-regulating techniques for electric motors, like frequency control and permanent magnetic motor speed control. Improve the flow control modes of fan and pump, eliminating mechanical throttling regulation methods. Reform the speed regulation methods of large–medium motors with variable operating conditions. Match the power between motors and loads reasonably.
3) Control of Driven Devices and Equipment Reform
   Replace the traditional mechanical drive system by an innovative power electronics drive system, and replace AC speed regulation by DC speed regulation. Reform driving equipments, especially including the large drainage and irrigation equipments and the enterprise with over 100,000 kW motor capacity.
4) Optimization of the Operation and Control for Electric Motor Systems
   Promote the application of soft starters, reactive power compensation devices, and computer-aided automatic control system. Fulfill the economic operation of the system by reasonable energy allocation in each process.
5) Key Reform Area
   The key reform areas include electric power, metallurgy, nonferrous metallurgy, coal, petroleum, chemical engineering, electromechanical, light industry, ventilation and air-conditioning system of enterprise, and electric motor system of building an air-conditioning system.

The industrial field has the greatest potential for the energy saving of electric motor system. But various working conditions, different load characteristics, and industrial processes make the energy saving of an electric motor system an extremely complicated project. For further study and application, energy saving of the electric motor system is divided into motor ontology energy saving, the drive and driven equipment system energy saving, pipe network system energy saving, reasonable match between different components and subsystems, and system control energy saving.

### 1.1.3 Status of Energy Saving for Electric Motors in China and Abroad

Energy saving of power not only means the reduction of energy used for electric power generation and the decrease of harmful gas effluence, like $CO_2$ and $SO_2$, but it also means the weakening of greenhouse effect; so energy saving for electric motors became a concerned issue in global range. In 1998, the United States proposed the "U.S. Department of Energy Motor Challenge Program." The research showed that efficiency improvement by reforming motors has 24.6 billion kWh per year energy-saving potential (4.3% of the total motor power consumption) and the efficiency improvement by reforming speed regulation ways has 60.6 billion kWh per year energy-saving potential (10.5% of the total motor power consumption); hence, the total energy-saving potential is 14.8%. Japan also took electric motor energy saving as an important issue and decided to decrease 10% power generation even though its unit gross national product energy consumption has already been the lowest in the world.

In Europe, European Union Directorate-General for Transport and Energy (EU-DGTE) decided to forbid the manufacturing and circulation of electric motors of Eff 3 grade. On November 11, 2007, EU set "Energy-Using Product" directive as the law implemented in EU member states, in which mandatory energy-saving indexes were regulated for electric motors. In 2009, EU regulated the minimum energy performance standards (MEPS) for electric motors, replacing IE1 grade motors with more efficient IE2 and IE3 grade motors.

Australia promulgated energy saving a mandatory standard for high-efficiency motors, which was also performed in New Zealand. Motors produced and imported in Australia and New Zealand must satisfy the lowest efficiency indexes regulated in the standard, which is close to Eff 1 of Europe and EPAct of the United States.

From the global view, the energy saving of electric motor systems started from the reform of the motor itself. With the development of internationalization of trade, energy saving of electric motors has become an international issue, and relative aspects should be unified, including unified standards, efficiency indexes, and test methods.

To unify the standards of energy saving of electric motors, the International ElectrotechnicalCommission (IEC) started the IEC/TC2 WG31 working group for the formulation of standard IEC60034-30 "Efficiency classes for single-speed electric motors." The standard applies to the motors of rated voltage 1000 V and below; rated power between 0.75 and 375 kW; number of poles are 2, 4, or 6; and motor duty S1 or S3 (duty cycle is 80% and above). It is stipulated that the standard does not apply to the motors studied in IEC60034-25 and the cage induction motor designed for power supply to variable-frequency drive.

The unification of efficiency indexes is the key issue for the standard formulation. IEC60034-30 divides the efficiency indexes into three grades, IE1, IE2, and IE3, applying to frequency systems, 50 and 60 Hz. In the three grades, IE1 is normal efficiency, IE2 is high efficiency, and IE3 is ultrahigh efficiency. For the 50 Hz system, energy efficiency IE1 equals to Eff 2 of EU and grade 3 of China, IE2 equals to Eff 1 of EU and grade 2 of China, and IE3 equals to grade 1 of China.

The premise of the unification of energy indexes is the unification of test methods. IEC60034-30 stipulates that efficiency test methods should refer to IEC60034-2-1 (2007), adopting middle and low uncertainty test methods for motors of IE1 and below, and adopting low uncertainty test methods for motors of IE2 and above. In IEC60034-2-1, the test method estimating stray loss by 0.5% of input power adopted in current energy efficiency standard of EU and China is abolished.

The current electric motor products series Y, Y2, Y3, YX of China adopts 0.5% of input power estimating stray loss. Chinese standard GB18613-2006 stipulates that from

January 2010, grade 2 which equals IE2 was adopted in China. The United States and Canada carried out IE2 in 1997, Mexico 1 year later in 1998, and Australia and New Zealand in 2006. In developed countries, IE3 has been carried out. Many countries formulated relevant laws and regulations to increase the market share of high-efficiency and ultrahigh-efficiency motors, in which the United States and Canada promoted the fastest (high-efficiency motors take 54% market share and ultrahigh-efficiency motors take 16% share), and China promoted relatively slowly (high-efficiency motors take only 1%). Now, the average efficiency of China is IE1 (average efficiency is 87%). According to medium- and long-term energy-saving plans of China, the motor efficiency level should be up to IE2 (average efficiency is around 90%) in 2010 and promoted the IE3 motors (average efficiency is around 92%). In 2017, about 50% of the motor efficiency level have been up to the IE3 standard. Thus, in China, various collaborations should be carried out between government, enterprises, and research departments in the areas including compulsory laws, law enforcement and supervision, market education, policy formulation, and product development.

## 1.2 Main Development Ways of Energy Saving for Electric Motor System

An electric motor system consists of power control devices, a motor, and a driven mechanism load. According to the statistics, half the power energy is consumed by electric motors, of which 80% is by a three-phase asynchronous motor in China. In global range, energy consumption of the electric motor is considerably great. So, energy saving of the electric motor system attracts high attention by many countries.

In the energy-saving plan during the 11th Five-Year Plan of China, electric motor energy saving received the top priority, and electric motor speed regulation technology and power electronics (power-saving technology) were proposed, accomplishing the economic operation of electric motor, fan, pumping equipment, and system. During the 11th Five-Year Plan of China, the energy saving aim was reduced to 20%. During the 12th Five-Year Plan of China, the energy saving aim was reduced to 16%. The energy saving of electric motor has both great potential and challenges.

The energy-saving principles and measures for various electric motors are different. Both purchase cost and operation cost are considered when choosing electric motors. High-efficiency motors have 20–30% of purchase cost than normal motors of the same series, but has lower operation cost since 97–98% of operation cost comes from the power consumed. So the most important aspect of energy saving for electric motors is the improvement of efficiency. The main development of energy saving for electric motors has the following three sections.

### 1.2.1 Efficiency Improvement of Y Series Asynchronous Motor

In various industries, a three-phase asynchronous motor consumes the most power energy. Until now, many efforts have been made to improve the efficiency of asynchronous motor. The key to improving efficiency is to reduce internal loss of the motor, and the main measures are as follows:

1) Increase effective materials to reduce winding loss and iron loss.
2) Adopt magnetic materials and technological measures to reduce iron loss.

3) Reform fan design to reduce ventilation loss.
4) Optimize the design and technological measures to reduce stray loss.
5) Optimize design by computer technology to reduce loss.
6) Reform die casting technology to reduce rotor losses, like adopting copper rotor.

New design and effective materials could reduce loss. A copper-winding conductor could reduce winding resistance and a silicon steel lamination could reduce the iron loss. But all the aforementioned measures will increase the manufacturing cost of motors.

### 1.2.2 Promoting Frequency Speed Regulation Technology

Frequency speed regulation technology mainly applies to a three-phase asynchronous motor or synchronous motor, and it is the ideal and most promising speed regulation methods. The frequency converter has both the excellent speed regulation performance and the remarkable energy-saving performance and thus becomes the ideal speed regulation device for manufactories. Frequency speed regulation technology could make motors work with the most energy saved speed by controlling rotating speed more precisely. Take wind turbines, for example. According to theory of fluid mechanics, shaft power is proportional to the cube of rotating speed. When the air volume needed and rotational speed are reduced, power decreases by cube of rotating speed, so the effects of frequency speed regulation are considerable. The capacity of similar motors whose loads change with rotating speed is set by the most need and this leads to more design margin. But in operation practice, motors usually work under light load condition. Adopting frequency speed regulation could improve the efficiency of the electric motor system greatly. So frequency speed regulation has great potential.

From 1980s, a frequency converter has been widely used in various industries for the speed control and energy saving of transmission systems.

### 1.2.3 Promoting High-Efficiency Motors and Permanent Magnet Motors

#### 1.2.3.1 High-Efficiency Electric Motor: An Important Way of Energy Saving

A high-efficiency asynchronous motor (YX series electric motor) could improve efficiency by 3% and reduce loss by 20–30% compared to a Y series motor. Billions of kWh power would be saved if high-efficiency asynchronous motors were applied in the whole of China. Investment for high-efficiency motors exceeds that for normal motors by 30% and could be recovered within 1–3 years. To accomplish the transformation from a normal motor to a high-efficiency motor, besides increasing the amount of silicon steel and copper wires and narrowing fan size, many efforts should be made on application of new materials, manufacturing technology, and design optimization to control costs and meet the limitation of structure size.

The main motors in market are Y and Y2 series products. The Y series products were designed in early 1980s, while the Y2 series products were designed in the mid-1990s, and 90% market share belongs to them. Though 40 or more industries could produce high-efficiency motors in China, like YX and GX series products, the market share of high-efficiency motors is just 10%.

High-efficiency motors emerged in 1970s, whereas ultrahigh-efficiency motors emerged in 1990s, internationally. Generally, high-efficiency motors have 20% less loss than normal motors and ultrahigh-efficiency motors have 30% less loss than normal motors. So high-efficiency motors and ultrahigh-efficiency motors have observable energy-saving effects on long-term and high load rate operating conditions.

#### 1.2.3.2 Permanent Magnetic Electric Motor: A New Kind of High-Efficiency Motor

The permanent magnetic electric motor, whose rotor is produced in permanent materials, could reduce 20–30% loss and increase 3–5% efficiency. It is more efficient than a high-efficiency motor. It has more advantages besides its high efficiency. First, it has smaller volume, just two-third of normal asynchronous motor with the same power and poles. Second, it has high starting torque, three times of rated torque generally. Third, it has high power factor, which could reach 0.95. Fourth, it could rotate slowly with high efficiency and power factor when the number of polar is less than 12. Lastly, it has better speed regulation performance, because high-performance permanent magnetic motor emerged with the development of power electronic technology and control technology.

The development of permanent magnetic motor has advantaged predominance. The main production material of a permanent magnetic motor is Nd Fe B permanent magnetic material which comes from rare earth. Seventy percent of the whole rare earth exists in China which lays a solid foundation for the development of the permanent magnetic motor. For all that, the main limitation of the development of permanent magnetic motor is still the cost of permanent magnetic materials.

## 1.3 Energy Saving: The Basic National Policy of China

On November 1, 1997, “Laws of Saving on Energy Resources in R. P. of China” was adopted at the Standing Committee of the Eighth National People’s Congress’s 28th meeting and its revised version was adopted at the Standing Committee of the 10th National People’s Congress’s 30th meeting on October 28, 2007 and carried out on April 1, 2008. The new energy-saving law placed energy saving as the basic national policy of China, and clearly points out: the nation performs the basic national policy of energy saving, and performs the energy development strategy of the implementation of conservation and development simultaneously, giving priority to conservation, and puts energy conservation as a long-term strategic approach to national economic development. The new energy-saving law increases from the original 6 chapters 50 provisions to 7 chapters 87 provisions. Compared with the original version, the new one further defines the energy-saving enforcement subject and strengthens the energy-saving legal liability. The sixth provision defines that the nation performs the energy-saving target responsibility system and energy-saving examination evaluation system, and takes the completion of energy saving as the evaluation aspects of local people’s governments and its principals.

The new energy-saving law adds legal weapon for the scientific development of China; it would help solve the increasingly acute conflict among Chinese economic development, energy resources, and the environment. The key in the implementation of energy-saving law lies on the Chinese government. The new energy-saving law strengthens the government's responsibility in terms of energy-saving management, but also makes regulations for government agencies to make their own energy conservation. The law revision strengthens the responsibility of not executing and not implementing the energy-saving law. The new energy-saving law has significant changes in the legal adjustment range and maneuverability; further standardizes energy-saving regulations of the industrial sector; and adds new energy-saving regulations in the field of construction, transportation, and public institutions for the current weakness of energy conservation.

To achieve the combination of market regulation and government management and play the role of economic instruments and market economy in the energy management, the new energy-saving law adds the incentive policies, specially including policies to promote the enterprise energy saving and industrial upgrading and mandatory measures to limit the development of high energy consumption and high pollution industries, like compulsory energy efficiency markers and elimination system. The new energy-saving law could help transform the weak energy-saving status, and ensure the completion of each stage of emission reduction targets in the legal level. For all these, energy saving is the basic national policy and long-term development strategy of China.

Corresponding policies have been put out in promoting high-efficiency motors. On June 2, 2010, China's Ministry of Finance and National Development and Reform Commission jointly promulgated "Information about printing and distributing implementation details of energy saving products, people-benefit project and high efficient energy saving motors," deciding to add high-efficiency energy-saving motors into energy-saving products and people-benefit project range and promote by financial subsidies. Information requires that manufactories are the promotion subjects of high-efficiency motors and sell products to customers at subsidized prices by central finance. At the same time, the promoting manufactories should sell no less than 300,000 kW of low voltage three-phase asynchronous motors and high voltage three-phase asynchronous motors, or sell no less than 10,000 kW of rare earth magnetic permanent motors.

Local governments also increased financial subsidies to promote a high-efficiency motor. Beijing municipal Development and Reform Commission planned to promote high-efficiency motors within 166 key energy consumption enterprises in 2 years and the total investment of the project is 230 million of which 90 million is from government finance.

Four energy policies will be adopted in the process of constructing an energy-saving society.

1) Through adjusting the industrial structure to promote energy conservation, China takes transforming the growth mode as the tactic stress point of economic progress and raises the quality and efficiency of economic development, giving full play to the energy-saving effectiveness of structure adjusting. Accelerating the optimization and upgrading of the industrial structure, enthusiastically implement the "provisional regulations of adjusting the industrial structure" and the "guide catalog of adjusting the industrial structure." Accelerate the development of energy-saving services and industries of high and new technology as well as control energy-intensive, material-intensive, and water-intensive industries strictly. According to the carrying capacity of resources and environment, optimize the local industrial distribution, and implement different policies of optimized exploitation, priority exploitation, limit exploitation, and prohibited exploitation.

2) Through improving price and taxation system to promote energy conservation, our country will insist on giving full play to the basic role of market forces in allocating resources, enthusiastically promoting the system reform in energy field; accelerate the market process of the energy product price; and establish the price system which can reflect the scarcity of the resources. Establish and improve the taxation policy system which benefits the construction of energy-saving society; and accelerate the taxation policy of encouraging production and usage of energy-saving products, including low fuel consumption and low-emission vehicles.
3) Through improving scientific and technological level to promote energy conservation, our country will insist on implementing policy of encouraging enterprises technology innovation; reinforce on absorption and reinnovation of the introduced technologies; implement self-innovation government procurement; implement tactic of intellectual properties and technical standards; implement taxation policy of encouraging innovative pioneering; accelerate the industrialization of high energy and new technology and the generalization of advanced technologies.
4) Through enhancing the management to promote energy conservation, our country will develop and revise the laws and regulations to promote effective use of resources; straiten the access of high energy-consuming and high polluting industries; accelerate to set compulsory standards of energy efficiency; develop and perfect the energy-saving design standards of the main energy-consuming industries; and establish the building energy-saving standards as soon as possible. Enhance power demand side management and promote contract energy management and voluntary energy-saving agreements; establish reward and punishment system and enlarge supervision of law enforcement.

Though energy intensity per GDP of China has decreased a lot since the promulgation of "Laws of Saving on Energy Resources in R. P. of China," great disparity still exists compared with developed countries. Great achievements have been obtained in the past 10 years in the field of energy saving of electric motors, but many relevant researches need to be done, like design theory of electric motor, manufacturing processes, motor materials, control technology, and drive power electronics technology.

## 1.4 Main Contents of This Book

Three-phase asynchronous motors are the most commonly used motors in industrial and agricultural areas and account for about 80% of the total installed capacity; so the book mainly focuses on energy-saving theory and technology of the three-phase asynchronous motor system.

Motor system converts electric power into mechanical power, consisting of power supply unit (power conversion unit), motor, and mechanical loads. Energy saving of motor system actually is the improvement of operation efficiency. Improving efficiency mainly relies on two ways: one is improving efficiency of the motor itself and the other is improving the match between motors and loads. The latter includes three aspects: power match, rotation speed match, and mechanical characteristics match.

The efficiency of motors usually means the rated efficiency, decided by design, material, manufacturing process, and production condition. In the book, the energy efficiency indexes and relevant standards are introduced briefly, and contents of design and manufacturing process could refer on relevant data.

Main contents of the book are the introductions of energy-saving theory and methods of three-phase asynchronous motors with examples. Essential problem of motors' energy saving is the improvement of efficiency. The economic operation of the three-phase asynchronous motors is introduced first, and then the power match, rotation speed match, and mechanical characteristic match between motors and loads. Besides, energy-saving theory and methods of reactive power compensation and energy-saving theory and application prospects of permanent magnetic motors are also introduced. Last, the test methods and evaluation methods of energy-saving effects are mentioned.

Relevant standards of China are described throughout the book for readers to grasp the standards besides the theories and technologies. Learn to grasp and use standards in test methods and evaluation methods.

# 2

# Overview of Three-Phase Asynchronous Motors

Three-phase asynchronous motors have been extensively used in various industrial applications, due to their advantages such as simple structure, high reliability, low cost, high overload capacity, and low maintenance. However, they also have some disadvantages of low power factor, large start-up current, and low starting torque. In order to further study the energy-saving characteristics of three-phase asynchronous motors, their basic structure, operating principle, power relationships, and mechanical characteristics are presented in this chapter. The energy efficiency standards of three-phase asynchronous motors for different countries are introduced in the chapter; especially the International Energy Efficiency Standard, and the allowable values and grades of energy efficiency of small and medium three-phase asynchronous motors in China. In addition, the main types and typical products of three-phase asynchronous motors, especially the characteristics and technical specifications of the Y2, Y3, and YX series energy-saving motors, are investigated and discussed. The purpose of the chapter is to let the readers understand the performance parameters and efficiency characteristics of various asynchronous motors as well as the allowable energy efficiency grades and product status of energy-saving motors, aiming at forming an important foundation for motor selection and energy-saving analysis.

## 2.1 Basic Structure and Characteristics of Three-Phase Asynchronous Motors

### 2.1.1 Basic Characteristics of Three-Phase Asynchronous Motors

AC motors mainly are of two types: asynchronous motor and synchronous motor. A three-phase asynchronous motor realizes energy conversion based on electromagnetic induction principle, so it can also be called "induction motors." As the rotor speed and stator magnetic field speed are different in most operating states, the induction motor is also called "asynchronous motor." In the national standard, it is collectively referred to the as three-phase asynchronous motor; thus, in this book they are all referred to as three-phase asynchronous motors or asynchronous motors. Asynchronous machines are widely used as driving devices in various industrial and agricultural

*Energy-saving Principles and Technologies for Induction Motors*, First Edition. Wenzhong Ma and Lianping Bai.

applications, such as machine tools, pump, mining equipment, and so on. The three-phase asynchronous machines have the following advantages:

1) Simple structure, high reliability, easy manufacturing, low cost, and long service life.
2) With the development of power electronic devices and variable frequency speed control technology, the performances of asynchronous motors with a wide speed range based on variable frequency speed control system are now comparable with DC motors.

When in operation, three-phase asynchronous motors must absorb reactive power from the grid, which would reduce the power factor of grid, especially worse at low speed. Unless using complex variable frequency speed control device, it is hard for asynchronous motors to economically achieve a continuous variable transmission in a wide speed range, at present. In addition, compared with the DC motors, the start-up current of three-phase asynchronous motors is large, but the starting torque is low, which brings some unfavorable factors in the selection and use of the asynchronous motors. Nevertheless, the asynchronous motors have the potentiality to achieve energy saving if they are used sensibly.

### 2.1.2 Basic Types of Three-Phase Asynchronous Motors

There are many methods to classify the three-phase asynchronous motors. According to the phase number of the power supply for stator winding, the asynchronous motors can be classified into single-phase and three-phase motors; according to the rotor structure, they can be classified into rotor winding and squirrel cage asynchronous motors; according to the working voltage, they can be classified into high voltage (6–10 kV) and low voltage (380 V) asynchronous motors; according to the operating performances, they also can be classified into high starting torque and high slip asynchronous motors; according to the power capacity they can be classified into small, medium, and large asynchronous motors. With reference to the national standard (GB 18613–2002), the power ranges of medium asynchronous motors is about 0.75–315 kW, large asynchronous motors is above 315 kW, and small asynchronous motors is below 0.75 kW. Thus the small-power motors can be also called fractional horsepower motors.

### 2.1.3 Basic Structure of Three-Phase Asynchronous Motors

Squirrel cage asynchronous motors have several distinctive advantages: simple structure, reliable operation, easy manufacture, and lower cost; the wound asynchronous machines generally have satisfactory performances on starting, braking, and speed control. At present, considering the fact that the squirrel cage three-phase asynchronous motors are the mainstream products, these motors can be taken as an example to introduce the basic structure of asynchronous motors. Squirrel cage three-phase asynchronous motors are made up of two parts: stator and rotor. The stator is the stationary part, while the rotor is the rotating part, and there is an air gap between them. Figure 2.1 is the basic structure of squirrel cage three-phase asynchronous motors.

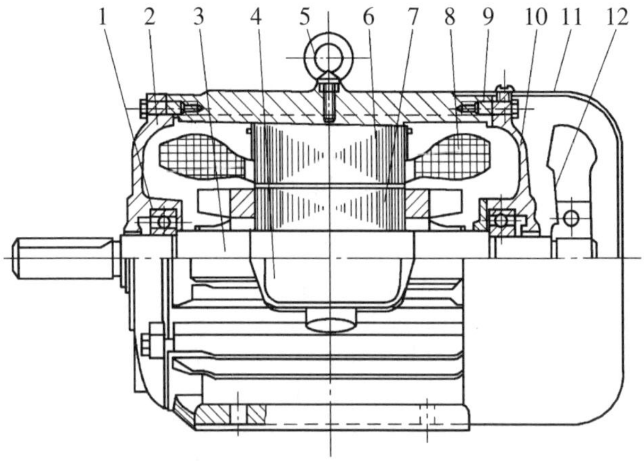


**Figure 2.1** Configuration of squirrel cage three-phase asynchronous motor. 1, bearing; 2, front end cover; 3, shaft; 4, terminal box; 5, ring; 6, stator core; 7, rotor; 8, stator winding; 9, frame; 10, back end cover; 11, fan housing; 12, fan.

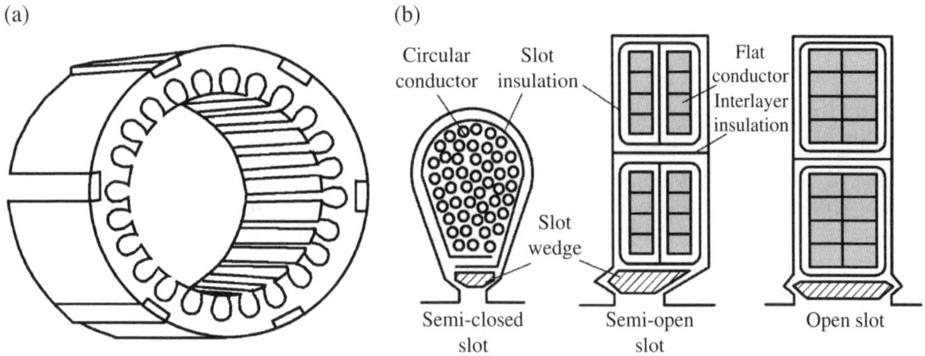


**Figure 2.2** (a) Stator core and (b) stator slot types.

#### 2.1.3.1 Stator

Asynchronous machines stator is composed mainly of following components: stator core, stator winding, frame, end cover, and so on.

The stator core is the part containing the main magnetic flux circuit of machines and the place where to lay in the stator winding, as shown in Figure 2.2a. In order to reduce the eddy current loss and magnetic hysteresis loss caused by a rotating magnetic field in the iron core, the stator core is formed by a lamination of 0.5 mm thick silicon steel sheet, with varnish coated on both sides of the silicon steel sheet to ensure insulating. Stator core is slotted concavely on the inner circle surface to allow the stator windings to be put in. The cross-sectional view of the slot may take the shape of a semi-closed slot, half open slot, and open slot, as shown in Figure 2.2b.

Stator winding is the circuit part of a motor, connected by a number of coils in a certain rule, and each coil has two effective edges in two slots, respectively. The function of stator winding is to establish a magnetic field, generate an electromotive force with current, and ultimately realize the electromechanical energy conversion. The arrangement of stator winding in groove can either be single-layer, or double-layer. The stator windings of three-phase asynchronous motors are usually composed of three-phase symmetric windings with identical winding setup, each winding as one phase, and therefore in a three-phase winding setup the spatial difference between any two neighboring phases is 120°. Usually two ends of the three-phase windings are expressed as U1–U2, V1–V2, W1–W2, and the six outlets are wired to the junction box, where they can configuratively be connected as Y or Δ type according to need.

The machine base is to fix and support stator iron core, so the requirement is that it must have enough mechanical strength and stiffness. Medium- and small-rated motors generally use a cast-iron frame, and the frame of big capacity asynchronous motors is welded by steel plates. According to the different cooling methods, different forms of frame are used. For the medium or small capacity asynchronous motors, there are often the radiating ribs on the frame surface to increase the cooling area. For the large capacity asynchronous motors, there is a separation space between the frame inner wall and stator core as a cooling passage. The role of the end cover is to install bearing to support the rotor so that the stator and the rotor can keep a certain degree of concentricity.

#### 2.1.3.2 Rotor

The rotor is composed of a core, the winding, and a turning shaft, and the rotor core is also a part of the magnetic circuit of the motors. In order to reduce the eddy current loss and the magnetic hysteresis loss, it is also made of a 0.5 mm thick silicon steel sheet. Overall, the rotor core, the air gap, and the stator core constitute the complete magnetic circuit of a motor, and the connection between the rotor core and the shaft must be reliably fixed to transfer mechanical power. On the outer circular surface of the rotor core is the inner groove with slots—rotor windings are laid in these slots. The rotor winding has several functions: induction of electromotive force, forming current path, and generating electromagnetic torque. The squirrel cage rotor winding formed by conducting bars and end rings is a symmetrically formed multiphase closed winding. Figure 2.3a shows a cast aluminum rotor which is formed by casting out together the conducting bar, the end rings, and the fan, making it the feature of simple structure and easy manufacturing and is often used in medium and small machines. Figure 2.3b shows

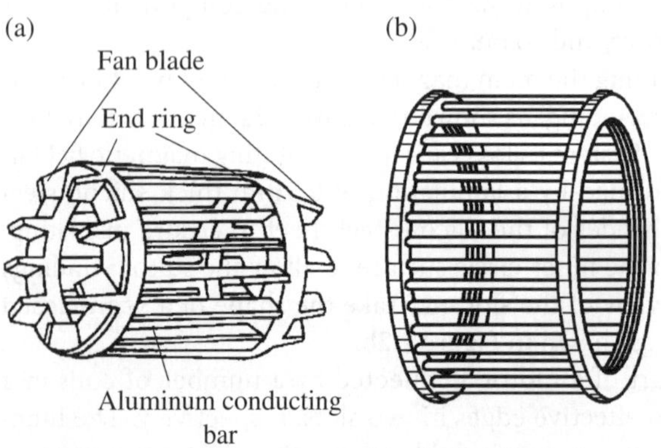


**Figure 2.3** The squirrel cage rotor. (a) Cast-aluminum rotor winding and (b) copper-cage rotor winding.

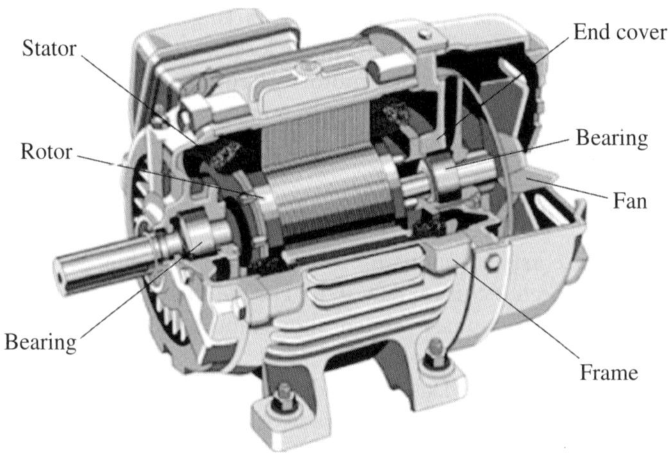


**Figure 2.4** Three-dimensional sectional drawing of a squirrel cage three-phase asynchronous machine.

a structure, often referred to as the inserting copper bar rotor, where all copper bars are welded to the end rings forming a short circuit winding. If the core is removed, the remaining part looks like a squirrel cage, hence the name squirrel cage rotor. For medium, small asynchronous motors, with a squirrel cage rotor, cast-aluminum technology is often used to cast out the workpiece containing the conducting bars, end rings, and fan blade altogether. A squirrel cage rotor does not need a current collector ring and has no requirement on insulation, so it has advantages of simple structure, easy manufacturing, low cost, and reliable operation.

In addition to stator and rotor, the squirrel cage three-phase asynchronous motors also have other parts like bearing, frame, fan, and so forth. Figure 2.4 is the 3D sectional drawing of a squirrel cage three-phase asynchronous motor.

Wound rotor winding, like stator winding, is also a symmetrical three-phase winding setup, often joined together into a star-connection configuration. The three outlets from the three windings are connected to the three slip rings on the shaft (called collector ring), respectively; the external circuit can be hooked up through a device called brush. In the circuit, there is usually an external additional resistor or a certain other device that connects in serial, through brush and slip rings, to the rotor loop, which can improve the performance of asynchronous motors on starting and speed regulation. Sometimes, the asynchronous motor is installed with a lifting brush short-circuit unit; this is used when the motor completes starting and does not need speed regulation, the operational handle lifts the brush causing the three slip rings to shorten at the same time to reduce the brush-wear loss and friction loss.

#### 2.1.3.3 Air Gap

The air gap size between stator and rotor has a great influence on the operating performance as well as the parameters of asynchronous motors. Due to the excitation current supplied by the power grid, the greater the air gap, the greater the excitation current required; besides it will also affect the power factor of the system. In order to improve the power factor of the motors, the air gap should be as small as possible; but if the gap is too small there will be difficulties in fabrication and assembly, and also bring about other adverse effects such as unreliable operation, poor starting, increasing high harmonic magnetic field which in turn increases the energy loss in conversion, and so forth. Generally, the air gap size of asynchronous motors often takes the minimum value that mechanical conditions allow; in general, the gap interval for medium and small capacity motors is 0.1–1.5 mm.

### 2.1.4 Basic Parameters of Three-Phase Asynchronous Motors

The nameplate of the three-phase asynchronous motors is shown in Table 2.1, with parameters of type, rating, and related technical data in it. The parameters that should appear on the nameplate of Y series three-phase asynchronous motors, according to the standard (JB/T 10391-2002), are as follows:

Motor name; Manufacturer; Motor type; Date of production; Winding connection; Insulation class; Duty type (working mode); Protection grade; Weight; Temperature rise. In addition to the above parameters, there are six rated parameters, namely rated power, rated speed, rated frequency, rated voltage, rated current, and rated power factor.

The rated parameters are defined as follows:

1) Rated power ($P_N$) refers to the output mechanical power on the shaft, when the motor is run in the nominal operating mode; unit in W or kW.
2) Rated voltage ($U_N$) refers to the line voltage exerted on the stator windings, when the motor is run in the nominal operating mode; unit in V.
3) Rated current ($I_N$) refers to the line current flowing through the stator windings, when the motor is run in the nominal voltage and rated power mode; unit in A.
4) Rated frequency ($f_N$) refers to the alternating frequency of mains; unit in Hz. (The frequency of mains in China is 50 Hz.)
5) Rated speed ($n_N$) refers to the rotor speed, when the motor is run in rated load; unit in r/min.
6) Rated power factor ($\cos\varphi_N$) refers to the power factor of each individual stator phase winding, which is determined by the phase winding voltage and the phase winding current, when the motor is run in nominal operating mode. For a three-phase asynchronous motor, the rated power can be calculated using the following formula:

$$P_N = \sqrt{3} U_N I_N \cos\varphi_N \eta_N$$

where $\eta_N$ is the rated efficiency, which is the ratio of output power to input power in nominal operation.

**Table 2.1** Nameplate of three-phase asynchronous motors.

| **Three-phase asynchronous motors** | | | | | |
|---|---|---|---|---|---|
| Type | Y2-225M-6 | Rated voltage | 380 V | Winding connection | Δ |
| Rated power | 30 kW | Rated current | 59.3 A | Working mode | $S_1$ |
| Rated speed | 980 r/m | Power factor | 0.84 | Temperature rise | 40°C |
| Rated frequency | 50 Hz | Insulation class | B | Protection grade | IP44 |
| ×××motors plant | | Date of production: | | Weight: 235 kg | |

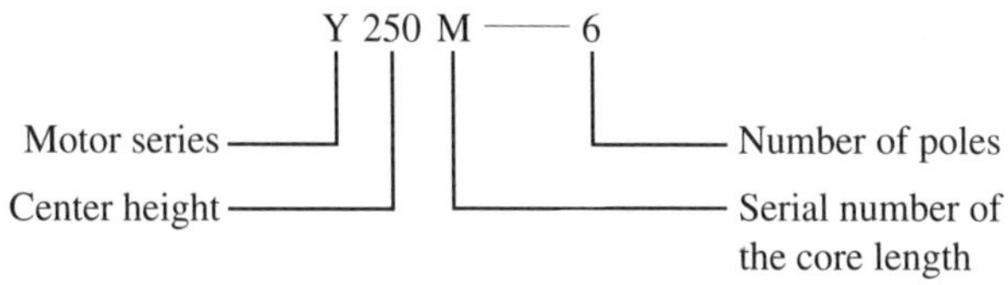


**Figure 2.5** Explanation of motor type on the nameplate.

In addition, the phase number of stator winding, connection method, efficiency, and other parameters are also marked on some nameplates. For wound rotor asynchronous motors, the rated voltage (refer to a rated voltage exerted on stator winding, or the line voltage between the rotor slip rings when the rotor is open) and the rotor rated current should also be marked.

The meaning of motor type on the nameplate is shown in Figure 2.5. Motor series Y represents a general series of squirrel cage asynchronous motors which are popular and mostly used. The length of the motor frame has three specifications labeled with S, M, and L representing the short, medium, and long frame sizes, respectively. Among M, L rating, sub-classification of M1, M2 or L1, L2, and so on can be made to designate a finer distinction. Frame length is essentially corresponding to the core length, and different frame lengths will yield different power when assuming the same center height. Other parameters in Figure 2.5 are 30 kW of motor power, M frame length, and 6 poles. Keeping other parameters fixed and altering the frame length, the motor power would be 22 kW if the length is S or 37 kW if length is L, respectively. When keeping both the center height and the frame length unchanged, the motor power would vary according to the number of poles.

## 2.2 The Principle of a Three-Phase Asynchronous Motor

The three-phase symmetrical windings are embedded in the stator iron of a asynchronous motor, and the rotor is a closed multiphase winding. Figure 2.6 shows the working principle of a three-phase asynchronous motor; Figure 2.6a denotes a stator winding

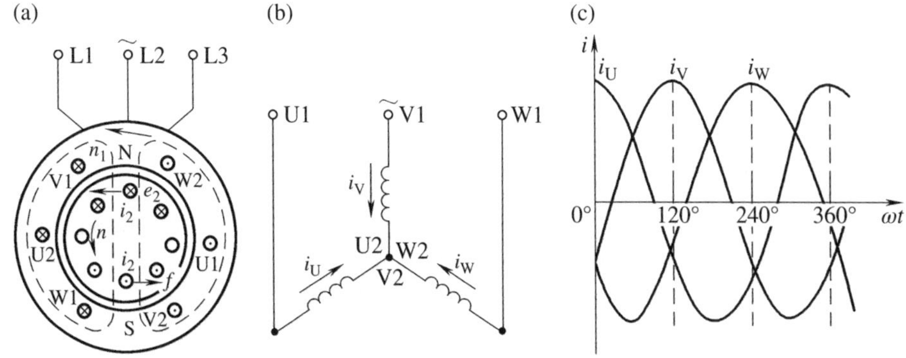


**Figure 2.6** The principle of three-phase asynchronous motors. (a) Stator winding space diagram, (b) connection diagram of three-phase stator windings, and (c) three-phase symmetrical currents.

space diagram for a three-phase squirrel cage asynchronous motor. There are three identical windings U1–U2, V1–V2, W1–W2 in the stator, and the spatial difference among any two of them is 120°. Such windings are called three-phase symmetrical windings. Conducting bars are put into the rotor slots, connected with the short-circuit ring at both ends, forming a closed cage winding. The small circles (Figure 2.6a) on the stator and rotor represent the windings of stator and the conducting bars of rotor. Figure 2.6b is the connection diagram of three-phase stator windings—either star-connection or triangle connection is allowed.

When the three-phase symmetrical stator windings are connected to a three-phase symmetrical power supply, the currents flowing through the three windings are also symmetrical. Suppose the instantaneous expression of each phase current is

$$i_{\mathrm{U}} = I_{\mathrm{m}} \cos \omega t$$
$$i_{\mathrm{V}} = I_{\mathrm{m}} \cos\left(\omega t - 120^{\circ}\right)$$
$$i_{\mathrm{W}} = I_{\mathrm{m}} \cos\left(\omega t - 240^{\circ}\right)$$

The plot of each phase current changing over time is shown in Figure 2.6c.

When the three-phase symmetrical currents flow through the stator of a three-phase asynchronous motor, a circularly rotating magnetic field is produced within the motor. As the concept of rotating field plays a very important role in motor theory, we give a detailed analysis on how it is formed and affected in terms of physics and structure.

We first study the three-phase winding with each phase consisting only one coil, as shown in Figure 2.7, in which U1–U2, V1–V2, and W1–W2 are the three coils allocated mutually with a special difference of 120°, distributed evenly on the inner circle surface of the stator iron core, forming a three-phase symmetrical winding. Such a three-phase symmetric winding possesses a pattern of phase displacement change in an anticlockwise manner, that is, phase V lags behind of phase U by 120° and phase W lags behind phase V by 120°.

Since the three-phase current changes continuously over time, and in order to easily study the composite magnetic field that is produced by three-phase symmetrical current, we observe the current values taken at several special times and learn how the magnetic field is changed accordingly. Let the four specially chosen times be $\omega t = 0$ ($t = 0$), $\omega t = 120°$ ($t = T/3$), $\omega t = 240°$ ($t = 2T/3$), and $\omega t = 360°$ ($t = T$). Assuming the current is positive, it refers to a current that is actually flowing out of the head end of the coil (U1, V1, W1) and in through the tail end of the coil (U2, V2, W2). A negative current refers to the current flowing out of the tail end and in through the head end of the coil. Sign ⊙ denotes a current flowing out from the paper and toward you, ⊗ denotes a current flowing into the paper away from you. First, look at the moment of $\omega t = 0$; by using either the instantaneous expression of current or the plot of current, we obtain $\omega t = 0$, $i_{\mathrm{U}} = I_{\mathrm{m}}$, $i_{\mathrm{V}} = i_{\mathrm{W}} = -I_{\mathrm{m}}/2$. If wet set the direction of each phase current on a diagram showing the cross-section of each phase winding, the current of phase U is positive and would flow out of U1 and in through U2, while the current of V and W would be negative and would flow in through V1 and W1 and out from V2 and W2. From this figure, we can conclude that current in the three coils of V2, U1, and W2 flows out from the paper with the current in V2 equal to the current in W2. According to the right hand rule, the resultant magnetic field of three coil currents would be centered with respect

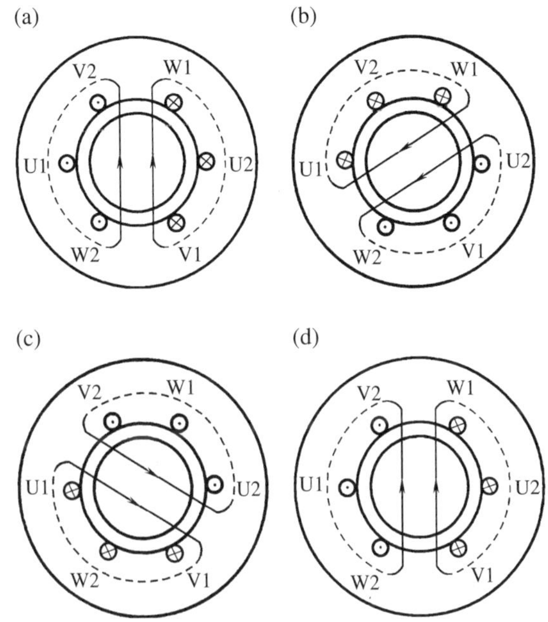


**Figure 2.7** Generation of a rotating circular magnetic field in motor. (a) $\omega t = 0°$, $i_U = I_m$, $i_V = i_W = -\frac{1}{2}I_m$; (b) $\omega t = 120°$, $i_V = I_m$, $i_W = i_U = -\frac{1}{2}I_m$; (c) $\omega t = 240°$, $i_W = I_m$, $i_U = i_V = -\frac{1}{2}I_m$; and (d) $\omega t = 360°$, $i_U = I_m$, $i_V = i_W = -\frac{1}{2}I_m$.

to the U1–U2 edge, but show an antisymmetrical distribution left and right due to the current directions in three coils. At this time when the field flows through the motor, the movement of field is upgoing from the bottom. Same reason stands for the analysis of the resultant field caused by three coil currents V1, U2, and W1, as shown in Figure 2.7a. The whole magnetic field is antisymmetrical (left–right) in the rotor. This feature of magnetic field direction (upgoing left–right) shows that the resultant magnetic field is only produced, as though, by a pair of magnetic poles. Similarly for the rest, the current direction and the magnetic field distribution for the other three chosen moments $\omega t = 120°$, $240°$, $360°$ can be similarly drawn out, as shown in Figure 2.7b, c, and d, respectively.

Examine Figure 2.7a–d; it can be seen that the composite magnetic field produced by the three-phase symmetric currents flowing in the three-phase symmetric windings is not stationary, and its direction is not alternating. It is like that the generated magnetic fields are, as though, produced by a pair of rotating magnetic fields, with the same magnitude of magnetic fields. From the moment of $\omega t = 0$ to the moment of 120°, 240°, 360°, the three-phase currents change in turn accordingly and the composite magnetic field produced by the three-phase currents turns 120°, 240°, 360°, respectively. The rotating direction is turning from phase U to V, then to W, that is, the rotating direction is $U \rightarrow V \rightarrow W$ (counterclockwise in figure). Based on the above discovery, it can be confirmed that the three-phase symmetric currents flowing in the three-phase symmetric windings would produce a rotating magnetic field with constant magnitude and speed.

Combining the current change and rotation of the rotating magnetic field shown in Figures 2.6 and 2.7, it clearly shows that when the three-phase currents change over a period $T$, the rotating magnetic field turns over 360° in space accordingly, namely, the current changes once of a period, the rotating magnetic field turns around a circle once.

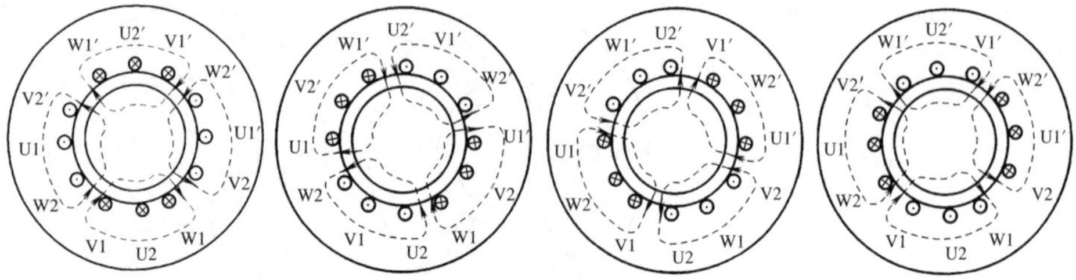


**Figure 2.8** Generation of motor's rotating field with two pairs of poles. (a) $\omega t = 0°$, (b) $\omega t = 120°$, (c) $\omega t = 240°$, and (d) $\omega t = 360°$.

Therefore, when the current changes $f_1$ (i.e., frequency) times per second, the rotating magnetic field will have turned around $f_1$ circles every second. Thus assuming, in a circumstance, that the rotating magnetic field has only one pair of poles, the relationship between the speed $n_1$(r/s) and frequency of alternating current is

$$n_1 = f_1$$

Figure 2.8 shows another arrangement of three-phase windings, where each phase of the three-phase windings U, V, and W consist of two coils (U1–U2, U1′–U2′), (V1–V2, V1′–V2′), and (W1–W2, W1′–W2′) connected in series, respectively. The span of each coil is 1/4 circle. It can be proved that for Figure 2.8, based on the aforementioned analysis method, the resultant magnetic field under the three-phase currents will remain as a rotating field. Here the pole number of the magnetic field is 4, that is, two pairs. As the current alternating once, the rotating field would turn 1/2 circle in this case. The concept can be extended by using 3, 4, ..., $p$ pairs of poles; a rotating magnetic field can be obtained as long as a three-phase winding structure and a three-phase current input is properly maintained. Applying the same method to study the relationship between the speed of the rotating field and the pole number of field, one can derive an inverse relationship between them, that is, as the current changes once of a cycle, the field with $p$ pairs of poles turns about $1/p$ circle. Thus for the mains frequency of $f_1$, the speed of rotating field with $p$ pairs of poles is

$$n_1 = \frac{60 f_1}{p}, \tag{2.1}$$

where $n_1$ represents the rotating field speed, and is also called the synchronous speed; its unit is r/min.

The working principle schematic diagram of asynchronous motors is depicted in Figure 2.9. Assume that the synchronous speed direction is counterclockwise, and if the rotor keeps static, the cage rotor bar would have a relative motion with the rotating field, and the direction of induction electromotive force is determined by the

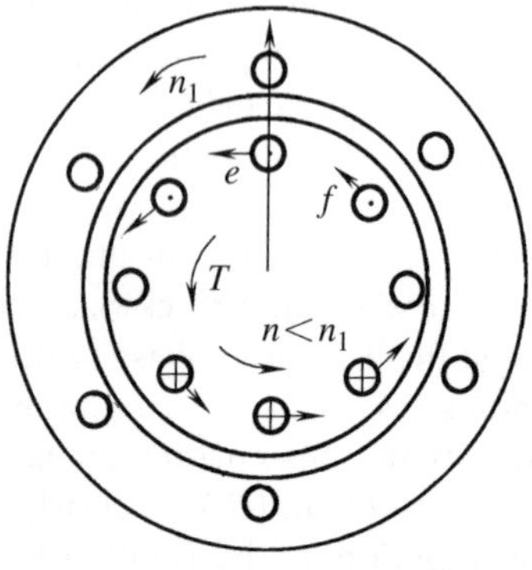


**Figure 2.9** Working principle schematic diagram of asynchronous motors.

right-hand rule. Since the rotor bars are linked together in their ends, the current would be formed in the bar, and we assume the current has the same direction with the electromotive force. So the force would put function on the bars in the magnetic field, whose direction is determined by the left-hand rule. Electromagnetic force forms an electromagnetic torque on the rotating shaft, and it is in the same direction with the rotating field. Under the influence of the electromagnetic torque, the rotor rotates with the rotating field, and the input electric energy is transformed into the rotating mechanical energy.

Above all, the basic principles of the three-phase asynchronous motors rotation are as follows:

1) Three-phase symmetrical current flows into three-phase symmetric windings to produce a circular rotating magnetic field.
2) Rotor conductor cutting magnetic rotating field produces induction electromotive force and current.
3) Electromagnetic torque is formed by the function of electromagnetic force on the current-carrying conductor of rotor, which forces the rotor to move.

The rotating direction of the asynchronous motors keeps the same with that of the rotating field, and the latter is determined by the phase sequence of the three-phase current, so the rotating direction of the asynchronous motors is consistent with the phase sequence of the three-phase current, which means if we change the phase sequence of the three-phase current, the motor would reverse.

Only if the speed of asynchronous motors is less than the speed of the rotating field, the motor would keep moving under the function of electromagnetic torque. If $n = n_1$, there will be no relative movement between rotor windings and stator magnetic field, then there would be no induction electromotive force and induction current in rotor windings. Therefore, $n < n_1$ is the necessary working condition for asynchronous motors, which is the reason known as "asynchrony." We called $\Delta n = n_1 - n$ as *slip*, and the ratio of $\Delta n$ and $n_1$ is called *fractional slip, simply as slip*, expressed as $s$.

$$s = \frac{n_1 - n}{n_1} \tag{2.2}$$

where $s$ is an important parameter of asynchronous motors and the implementation of $s$ would simplify the calculation than that of $n$. Usually, the rated $s$ of asynchronous motors is between 0.02 and 0.05; if the asynchronous motor stay static, $n = 0$, then $s = 1$. If the rotor's speed is close to synchronous speed (no-load operation), $n \approx n_1$, then $s \approx 0$. In general, for asynchronous motors, its speed ranges from 0 to $n_1$ and the slip $s$ ranges from 0 to 1. The increase of load would slow down the motor, so the value of $s$ reflects the speed of the motor and the size of loads. To improve the efficiency of common asynchronous motors, the rated slip is usually very small, set as 0.02 around.

## 2.3 Working Characteristic of Three-Phase Asynchronous Motors

To analyze the working characteristic of three-phase asynchronous motors, we need to introduce the equivalent circuit of the three-phase asynchronous motor first.

### 2.3.1 Equivalent Circuit of Asynchronous Motors

#### 2.3.1.1 *T* Type Equivalent Circuit of Asynchronous Motor

*T* type equivalent circuit of asynchronous motor is shown in Figure 2.10.

In the equivalent circuit, $r_1$ stands for the stator resistance of each phase, $x_{1\sigma}$ stands for the stator leakage reactance, which corresponds to stator flux leakage reactance, $Z_1 = r_1 + jx_{1\sigma}$, $Z_1$ stands for the leakage impedance of stator winding; $r_2'$ stands for the equivalent rotator resistance of each phase converted to stator side, $x_{2\sigma}'$ stands for the equivalent rotator reactance converted to stator side; $r_m$ stands for excitation resistance, reflecting the equivalent resistance of iron loss, and $x_m$ stands for the excitation reactance, which corresponds to the main flux; $Z_m = r_m + jx_m$, $Z_m$ stands for excitation impedance. *T* type equivalent circuit can reflect the electromagnetic law of asynchronous motors precisely. When the motor operates in the no-load condition, $n \approx n_1$, $s \approx 0$, $(1-s)/s$, $r_2' \to \infty$, $I_2 \approx 0$, the rotor circuit can be regarded as open-circuit. Then we can get the no-load circuit as shown in Figure 2.11a. The $x_m$ has a very big value, so the motors' power factor $\cos\varphi_o$ is very low.

In the condition of start, $n = 0, s = 1, ((1-s)/s)r_2' = 0$, the excitation branch in Figure 2.10 can be regarded as disconnected, so we can get the equivalent circuit of the start operation as shown in Figure 2.11b. $r_k = r_1 + r_2'$ and $x_k = x_{1\sigma} + x_{2\sigma}'$, stand for short-circuit resistance and reactance of asynchronous motors, respectively.

When in the rated condition, $s_N < 5\%$, $r_2'/s$ is more than 20 times larger than $r_2'$. So the rotor side branch of *T* equivalent circuit in Figure 2.11 presents resistance property, which improves the overall power factor of stator side greatly to 80–85%.

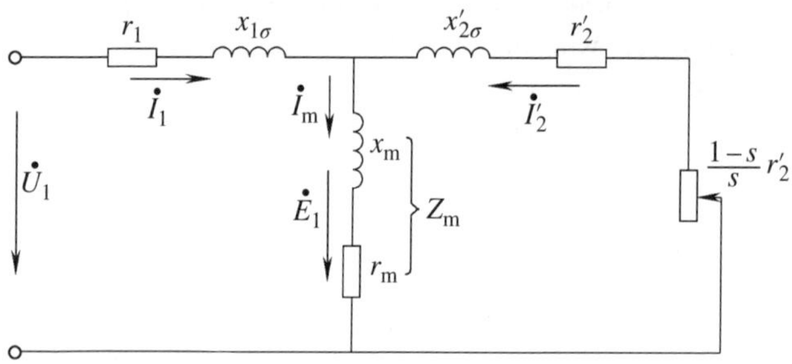


Figure 2.10 *T* equivalent circuit of asynchronous motors.

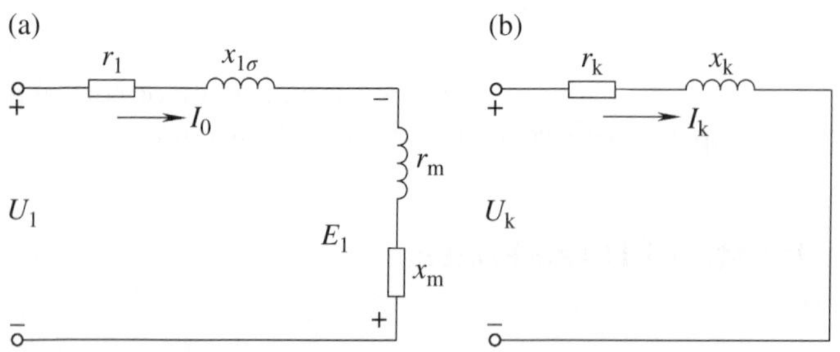


Figure 2.11 (a) No-load equivalent circuit of asynchronous motors and (b) starting or locked-rotor equivalent circuit of asynchronous motors.

Through the $T$ type equivalent circuit, because of the small value of $Z_1$, the voltage drop of stator leakage impedance $I_1Z_1$ changes very little from no-load to rated load. So we can regard $E_1 \approx U_1$ as a constant, and as a result, the main flux $\Phi_m$ stays nearly the same, so does $I_m \approx I_o$.

#### 2.3.1.2 Simplified Equivalent Circuit of Asynchronous Motors

$T$ type equivalent circuit is a series-parallel hybrid circuit, which results in the complex computing. To simplify the calculation, we usually put the excitation branch $Z_m$ to the input side to get a simplified equivalent circuit of asynchronous motors similar to $\Gamma$ type equivalent transformer circuit as shown in Figure 2.12. Because $Z_m$ is large and $Z_1$ is small, the error for motor caused by removing $Z_m$ branch forward is larger than that for transformer. Therefore, the series branch $Z_1 = r_1 + jx_{1\sigma}$ is added to the excitation branch in Figure 2.12 to reduce the error.

Besides the calculation simplification, the parameters $Z_k = r_k + jx_k$ and $Z_m = r_m + jx_m$ of the simplified equivalent circuit can be measured through experiment, which makes it a practical equivalent circuit in project.

No matter which equivalent circuit is adopted, we must pay attention that all parameters of stator and rotor in the equivalent are of one single phase, and all parameters of the rotor side are converted to stator side.

### 2.3.2 Power Balance of Asynchronous Motors

Power conversion process is the base to study the energy-saving property of an asynchronous motor. The electric power is input through the stator windings and mechanical power is output through rotor windings. The electromagnetic power is transited from the stator to the rotor by the air gap.

When operating, stator absorbs electrical power from grid and rotor outputs mechanical power to the dragging mechanical load. In the process of electromechanical energy conversion, there would be power loss inevitably. According to the law of conservation of energy, the output power equals the input power minus the total power losses.

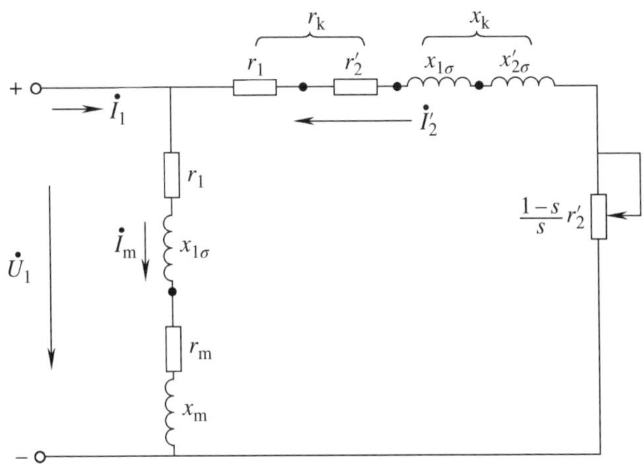


**Figure 2.12** $\Gamma$ equivalent circuit of asynchronous motors.

The power supplied to the motors from grid is called input power, and its computation formula is

$$P_1 = m_1 U_1 I_1 \cos\varphi_1 \tag{2.3}$$

When stator current flows through stator winding, the power loss generated by current $I_1$ on the stator winding resistance $r_1$ is called stator copper loss, and its calculation formula is

$$P_{\mathrm{Cu1}} = m_1 r_1 I_1^2 \tag{2.4}$$

Rotating field will produce iron loss in stator core (the rotor frequency is very low, about 1–3 Hz in general, and the rotor is formed by overlying, so the rotor iron loss is small, which can be neglected), its value can be regarded as the consumed power generated by current $I_o$ on the excitation resistance.

$$P_{\mathrm{Fe}} = m_1 r_{\mathrm{m}} I_0^2 \tag{2.5}$$

So deducting the stator copper loss $P_{\mathrm{Cu1}}$ and the stator iron loss $P_{\mathrm{Fe}}$ from the input power $P_1$, the remaining power is the electromagnetic power $P_{\mathrm{em}}$ from stator side to rotor side through the air gap field, that is

$$P_{\mathrm{em}} = P_1 - (p_{\mathrm{Cu1}} + p_{\mathrm{Fe}}) \tag{2.6}$$

According to $T$ type equivalent circuit, the electromagnetic power $P_{\mathrm{em}}$ entering into rotor loop is

$$P_{\mathrm{em}} = m_1 E_2' I_2' \cos\varphi_2 = m_1 \dot{I}_2'^2 \frac{r_2'}{s} = m_1 \dot{I}_2'^2 r_2' + m_1 \dot{I}_2'^2 \left(\frac{1-s}{s} r_2'\right) = P_{\mathrm{Cu2}} + P_\Omega \tag{2.7}$$

So part of $P_{\mathrm{em}}$ is consumed on the rotor winding resistance, $p_{\mathrm{Cu2}} = m_1 r_2' I_2'^2$; the remaining power is the mechanical power $P_\Omega$ on the shaft, which is the electrical power that consumed on the addition resistance $((1-s)/s)r_2'$ by the rotor current according to the equivalent circuit. That is,

$$P_\Omega = P_{\mathrm{em}} - P_{\mathrm{Cu2}} = m_1 \frac{1-s}{s} r_2' I_2'^2 \tag{2.8}$$

as shown in Figure 2.13.

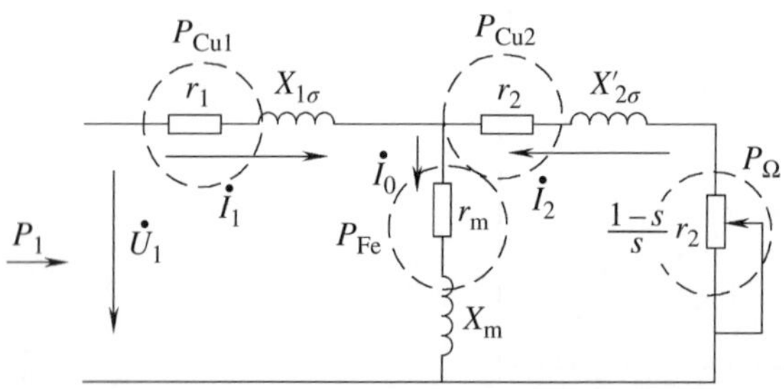


**Figure 2.13** Schematic diagram of equivalent circuit's loss of asynchronous motors.

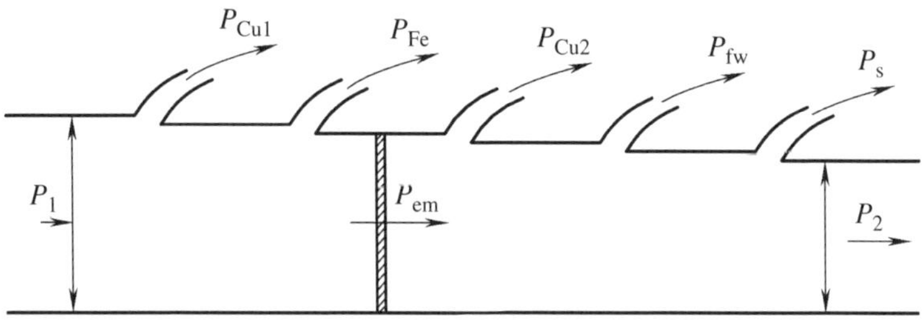


**Figure 2.14** Power flowing diagram of asynchronous motors.

According to Equations (2.7) and (2.8), we obtain

$$P_{Cu2} = m_1 \dot{I}_2'^2 r_2' = sP_{em} \tag{2.9}$$

$$P_\Omega = m_1 I_2'^2 \left(\frac{1-s}{s} r_2'\right) = \frac{1-s}{s} P_{Cu2} = P_{em} - P_{Cu2} = (1-s)P_{em} \tag{2.10}$$

According to the two equations, part of the electromagnetic power delivered from stator to rotor side through the air gap becomes the rotor copper loss $sP_{em}$, the remaining becomes the total mechanical power $(1-s)P_{em}$. If $n_1$ and $P_{em}$ keep constant, when $n$ decreases, namely $s$ increases, the output power would decrease, while the copper loss of rotor circuit ($sP_{em}$ is also called slip power) would increase which lowers the efficiency.

While the total mechanical power $P_\Omega$ on shaft is comprised of mechanical loss $P_{fw}$, additional loss $P_s$ and the output mechanical power $P_2$.

$$P_2 = P_\Omega - P_{fw} - P_s = P_1 - P_{Cu1} - P_{Fe} - P_{fw} - P_s - P_{Cu2} = P_1 - \sum P \tag{2.11}$$

where $\sum P = P_{Cu1} + P_{Fe} + P_{fw} + P_s + P_{Cu2}$ is the total power loss of the asynchronous motors.

Then the efficiency of asynchronous motors is

$$\eta = \frac{P_2}{P_1} \times 100\% = \frac{P_1 - \sum P}{P_1} \times 100\% = \left(1 - \frac{\sum P}{P_1}\right) \times 100\% \tag{2.12}$$

In conclusion, the power flowing diagram of asynchronous motors is shown in Figure 2.14.

### 2.3.3 Working Characteristics of Three-Phase Asynchronous Motors

The working characteristic of an asynchronous motor is the relationship between the output power and the motor's speed $n$, stator current $I_1$, electromagnetic torque $T$, power factor $\cos\varphi_1$, and efficiency $\eta$ from no load to rated load under the condition of rated voltage, rated frequency, rated connection, and no external impedance-stator/rotor loop. We can get the working characteristic of an asynchronous motor by calculation method. With the knowledge of equivalent circuit parameters, mechanical loss, and additional loss, $n$, $I_1$, $T$, $\cos\varphi_1$, and $\eta$ can be calculated; then we can get the working characteristics of the motor. We can also get the working characteristic by experimenting for a finishing motor. For middle and small types,

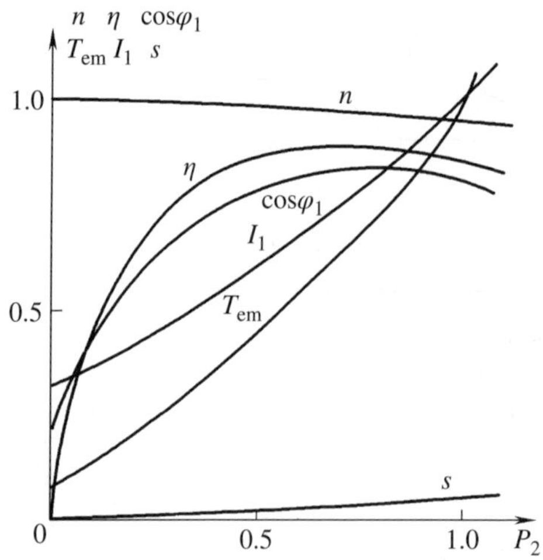


**Figure 2.15** Working characteristic of asynchronous motor.

using dynamometer as load, the output torque $T_2$ and $I_1$, $\cos\varphi_1$ can be measured under different speed, then the $P_2$, $\eta$ can be calculated and the characteristic can be got. For large-capacity asynchronous motors, parameters are usually obtained through no-load and short-circuit tests and then the working characteristic can be calculated by an equivalent circuit. The working characteristics of three-phase asynchronous motors are depicted in Figure 2.15.

#### 2.3.3.1 Speed Characteristic

The speed characteristic is shown in Figure 2.15 by curve $n$ or curve $s$. It can be seen that when $P_2=0$, $n\approx n_1$, $s\approx 0$. The increase of output power $P_2=T_2\Omega$ would cause the increase of the load torque ($T_2$), and to keep the balance of the electromagnetic torque and load torque, rotor current must increase, so the relative cutting speed $\Delta n=n_1-n=s_{n1}$ between air gap magnetic field and rotor must increase. The slip of asynchronous motors at rated speed is generally very small, ranging from 0.015 to 0.05. From no-load to full-load, the speed drops little from $n_1$ to $n_N=(1-s_N)\ n_1=(0.985-0.95)n_1$, so the speed characteristic $n=f(P_2)$ drops very little in the process.

#### 2.3.3.2 Stator Current Characteristic

The stator current of motors is $\dot{I}_1=\dot{I}_m+(-\dot{I}_2')$. In the operation condition of no-load, $I_1\approx I_m$, the $s$ and $I_2'$ would increase with the increase of load $P_2$, which would increase the load component of stator current. So $I_1$ is almost proportional of $P_2$. Characteristic curve of stator current is shown as curve $I_1$ in Figure 2.15.

#### 2.3.3.3 Electromagnetic Torque Characteristic $T=f(P_2)$

Motor's electromagnetic torque is $T=T_2+T_0$. From no-load to full-load, its speed $n$ and main magnetic flux $\Phi_m$ nearly remain unchanged, so does the mechanical loss $P_{fw}$, additional loss $P_s$, and the inherent no-load torque $T_0$. Because output torque $T_2=P_2/\Omega$ nearly increases proportionally with $P_2$, $T=f(P_2)$ is approximate to a rising straight line.

#### 2.3.3.4 Stator Power Factor Characteristic

It can be seen from the equivalent circuit of the motors that the circuit is inductive, so the lagging power factor of asynchronous motors is always below 1. In the condition of no-load, $\dot{I} \approx \dot{I}_m$ is basically reactive current, so $\cos\varphi_{10} < 0.1$. When the load increases, the active component of stator current would increase, and the $\cos\varphi_1$ would increase. Usually motors are designed to achieve the highest $\cos\varphi_N$ in rated load. In condition of overload, if $P_2$ continues to increase, the slip $s$ becomes larger, along with the increase of the impedance angle of rotor circuit $\varphi_2 = \arctan(sx_{2\sigma}/r_2)$, the decrease of $\cos\varphi_2$ and $\cos\varphi_1$, as shown in the curve $\cos\varphi_1$.

#### 2.3.3.5 Efficiency Characteristic $\eta = f(P_2)$

According to the efficiency formula of asynchronous motors,

$$\eta = \frac{P_2}{P_1} = \frac{P_2}{P_2 + (P_{fw} + P_s + P_{Fe}) + (P_{Cu1} + P_{Cu2})}.$$

We can get the curve of $\eta$ as shown in Figure 2.15. During the process from no-load to full-load, the change of main magnetic flux is very small, so the iron loss basically remains unchanged; the speed changes very little, so the mechanical loss also can be regarded as constant. Therefore, $(P_{fw} + P_s + P_{Fe})$ can be considered as constant, called fixed loss, while the copper loss of the rotor and the stator are proportional to the square of their current, respectively, called variable loss. In the condition of no-load, $P_2 = 0$ and $\eta = 0$, with the increase of $P_2$, the $\eta$ increases too and when the increase of $P_2$ makes variable loss equals to fixed loss, the efficiency would reach to its highest. Generally motors are designed to achieve the highest efficiency in rated load $P_N$. If load continues increasing, since the increase of $(P_{Cu1} + P_{Cu2})$ is proportional to the square of current, the denominator of $\eta$ expression grows faster than molecular, which would lower the efficiency.

Because the $\cos\varphi_1$ and $\eta$ of asynchronous motors both reach to maximum value around the condition of rated load, we usually make the motor operate under the rated condition during most time when choosing the capacity of the motors. Low $\eta$ and $\cos\varphi_1$ go against the energy-saving property of motors.

## 2.4 Mechanical Characteristics of Three-Phase Asynchronous Motors

Mechanical characteristics refer to the functional relationship $n = f(T)$ between speed and electromagnetic torque of the motors under the condition of fixed motors' stator voltage $U_1$, power frequency $f_1$, and motors' parameters. Because there is a linear relation between slip and the speed, $s = f(T)$ can also be used to express the mechanical characteristics of three-phase asynchronous motors.

### 2.4.1 Three Types of Formulas of Mechanical Characteristics

#### 2.4.1.1 Physical Formula of Mechanical Characteristics

Physical formula of the motors' electromagnetic torque is

$$T = \frac{P_{em}}{\Omega_1} = \frac{pm_1N_1k_{\omega 1}}{\sqrt{2}}\Phi_m I_2' \cos\varphi_2 = C_{M1}\Phi_m I_2' \cos\varphi_2 \tag{2.13}$$

where $C_{M1}$, the torque constant of the asynchronous motors converted to stator side, $C_{M1} = (pm_1N_1k_{\omega 1})/\sqrt{2}$;
$\Phi_m$, the per pole flux of fundamental wave magnetic field;
$I_2'$, the conversion value of rotor current;
$\cos\varphi_2$, the power factor of rotor loop.

This formula is derived based on physical concept, which is exactly the same with electromagnetic torque formula derived according to $T = \sum f(DE2)$, so the formula (2.13) is called physical expression of the mechanical characteristic of asynchronous motors.

Physical formula reflects the physical generation essence of asynchronous motors' electromagnetic torque, which is the interaction between the main magnetic flux $\Phi_m$ and the rotor current active component $I_2' \cos\varphi_2$. It is the specific performance of the electromagnetic force law in the asynchronous motors.

#### 2.4.1.2 Parameter Formula of Mechanical Characteristic

From the equivalent circuit of motors, we can conclude the parameter formula of the motors' electromagnetic torque $T$ as follows:

$$T = \frac{m_1}{\Omega_1}\frac{U_1^2 r_2'/s}{\left(r_1 + \frac{r_2'}{s}\right)^2 + (x_{1\sigma} + x_{2\sigma}')^2} = \frac{pm_1U_1^2 r_2'/s}{2\pi f_1\left[\left(r_1 + \frac{r_2'}{s}\right)^2 + (x_{1\sigma} + x_{2\sigma}')^2\right]} \tag{2.14}$$

Expression (2.14) is composed of asynchronous motors' power source $U_1$, frequency $f_1$, parameters ($r_1$, $r_2'$, $x_{1\sigma}'$, $x_{2\sigma}$, $p$, and $m_1$) and slip $s$, so it is called parameter formula of mechanical characteristic. Obviously, when $U_1$, $f_1$ and motor parameters stay unchanged, electromagnetic torque $T$ is only related to slip $s$. Calculating the values of $T$ corresponding to $s$, which ranges from $-\infty$ to $\infty$, we can get the curve $T=f(s)$, as shown in Figure 2.16. After simple coordinate transformations based on the relationship between $s$

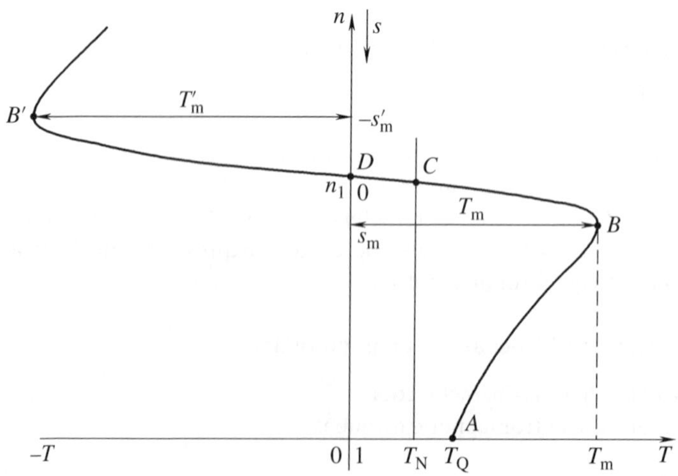


**Figure 2.16** Mechanical characteristic of three-phase asynchronous motors.

and $n$, the curve $T=f(s)$ could be transformed to the mechanical characteristics of asynchronous motors $n=f(T)$.

In Figure 2.16, the slip $s_m$ corresponding to the maximum electromagnetic torque $T_m$ is called critical slip.

The critical slip formula is

$$s_m = \pm \frac{r_2'}{\sqrt{r_1^2 + \left(X_{1\sigma} + X_{2\sigma}'\right)^2}} \tag{2.15}$$

Plugging $s_m$ in Equation (2.14), a maximal torque can be obtained as follows:

$$\begin{aligned} T_m &= \pm \frac{m_1}{\Omega_1} \frac{U_1^2}{2\left[\pm r_1 + \sqrt{r_1^2 + \left(x_{1\sigma} + x_{2\sigma}'\right)^2}\right]} \\ &= \frac{p m_1 U_1^2}{4\pi f_1 \left[\pm r_1 + \sqrt{r_1^2 + \left(x_{1\sigma} + x_{2\sigma}'\right)^2}\right]} \end{aligned} \tag{2.16}$$

Plus signify asynchronous motors operating in electric state in the first quadrant. Minus signifies operating in power-state in the second quadrant.

For asynchronous motors, usually $r_1$ is far less than $(x_{1\sigma}+x_{2\sigma}')$, the two formulas above can approximate to

$$\left.\begin{aligned} s_m &\approx \pm \frac{r_2'}{x_{1\sigma} + x_{2\sigma}'} \\ T_m &\approx \pm \frac{p m_1 U_1^2}{4\pi f_1 \left(x_{1\sigma} + x_{2\sigma}'\right)} \end{aligned}\right\} \tag{2.17}$$

Similarly, among them, " + " is used for electrical state and "–" stands for generating state. By formulas (2.16) and (2.17), $|s_m|$ of the electric and power state are the same. But because of the existence of $r_1$, $|T_m|$ of power state is larger than that of electric state. Ignoring the influence of $r_1$, $|T_m|$ of electric status is equal to that of power status. For the sake of simplicity, it can be thought that $T_m$ of power state equals to that of electric state in the later analysis and calculation.

$T_m$ is the maximum torque that asynchronous motors may produce. To avoid the motors being stopped by short-time overload in the running, it is required that rated torque $T_N < T_m$, making motors have a certain overload ability. We call the ratio of maximum torque $T_m$ to the rated torque $T_N$ as overload capacity, expressed as $\lambda_M = T_m/T_N$. $\lambda_M$ is an important indicator of asynchronous motors, and generally $\lambda_M = 1.6–2.2$.

The torque formula shows that the maximum torque is mainly related to the following factors:

1) Under the condition that power frequency and motor's parameters are invariable, the maximum torque is proportional to square of the stator phase voltage, and the critical slip has nothing to do with $U_1$, remaining the same.

2) If we keep $U_1, f_1$, and other parameters unchangeable and just change the resistance $r_2'$ of rotor loop, the maximum torque keep constant but critical slip changes, which is proportional to $r_2'$.
3) If we keep $U_1, f_1$, and other parameters unchangeable and just change $(x_{1\sigma}+x_{2\sigma}')$, $s_m$ and $T_m$ are inversely proportional to $(x_{1\sigma}+x_{2\sigma}')$ approximately.

When motor starts, $s=1$ is plugged in (2.14) to gain the parameter expression of starting torque of asynchronous motors:

$$T_Q = \frac{pm_1U_1^2r_2'}{2\pi f_1\left[\left(r_1+r_2'\right)^2+\left(X_{1\sigma}+X_{2\sigma}'\right)^2\right]} \tag{2.18}$$

According to expression (2.18), the starting torque is proportional to the square of power supply voltage $U_1$.

Inherent starting torque $T_Q$ of asynchronous motors can be obtained under conditions of rated voltage, rated frequency, and inherent parameters. The ratio of rated torque to starting torque is called starting torque ratio of asynchronous motors, expressed as $K_M = T_Q/T_N$. Generally, $K_M > 1$ is required so that motors can start smoothly driving rated load.

Three-phase asynchronous motors' parameter expression is commonly used for analyzing the influence of the voltage, frequency, and structure parameters on the mechanical characteristics.

#### 2.4.1.3 Practical Expression of Mechanical Characteristic

Calculating mechanical characteristics of three-phase asynchronous motors by parameter formulas must be based on prior knowledge of parameters, which are usually hard to obtain. If we can utilize motors' nameplate data and rated value provided by relevant manuals to calculate, it will become practical and convenient. The practical expressions of asynchronous motors' mechanical characteristic is

$$T = \frac{2T_m}{\dfrac{s}{s_m}+\dfrac{s_m}{s}} \tag{2.19}$$

Formula (2.19) is very convenient and practical, so it is called practical expression of mechanical characteristic. $T_m$ and $s_m$ in this expression can be calculated through following formulas:

$$T_m = \lambda_M T_N \tag{2.20}$$

$$T_N = \frac{9550P_N}{n_N} \tag{2.21}$$

The units of $P_N$, $n_N$, $T$ in the expression are kW, r/min and Nm respectively. $T=T_N$, $s=s_N$ are substituted into expression (2.20), and we can obtain the following:

$$s_m = s_N\left(\lambda_M+\sqrt{\lambda_M^2-1}\right) \tag{2.22}$$

Therefore, as long as $P_N$, $n_N$, and $\lambda_M$ of the motors are found out in motors' product catalogue, $T_m$ and $s_m$ can be calculated by formulas (2.20)–(2.22) and then plugged in formula (2.19) to obtain the practical expression of mechanical characteristic of the motors.

When three-phase asynchronous motors operate around rated loads, slip is very small and the rated slip $s_N$ just ranges in 0.015–0.05. According to Figure 2.16, when $s < s_m$, the mechanical characteristic is a straight line segment, while when $s = s_m$, $s/s_m = s_m/s$, the formula (2.19) can be simplified to

$$T = \frac{2T_m}{s_m} s \tag{2.23}$$

This is the expression of mechanical characteristic in a straight line segment and it is also known as linear expression of mechanical characteristic, which is similar to the characteristics of separately excited DC motors. The calculation method of $T_m = \lambda_N T_N$ is still calculated by Equation (2.20), and as for $s_m$, it needs to apply the new formula. Plugging $T = T_N$, $s = s_N$ into formula (2.20), we can obtain the following:

$$s_m = 2\lambda_M s_N \tag{2.24}$$

Linear expression (2.24) is easier and more convenient, but two points must be paid attention to:

1) This formula can only be used when operation is in the straight line segments of mechanical characteristics or $s = s_m$. If we cannot determine the operating condition, we can only use either practical expression or parameter expression.
2) The $s_m$ in the linear expression must be solved by $s_m = 2\lambda_M s_N$, instead of expression (2.22).

### 2.4.2 Inherent Mechanical Characteristic of Asynchronous Motors

Inherent mechanical characteristic curve is the measured relationship between $n$ and $T$, $n = f(T)$ of asynchronous motors in the condition that the three-phase stator windings of the motors are directly supplied to symmetrical three-phase voltage of rated voltage, rated frequency with regulated connection mode without any external impedance (resistance or reactance), and the rotor loop is directly shorted out without any impedance connected, as shown in Figure 2.16. There are several special operating points:

1) Starting point $A$: $s = 1$, $n = 0$, the corresponding electromagnetic torque of this point is inherent starting torque $T_Q$, also the starting torque at the time of direct start; the corresponding stator current is the starting current $I_Q$ at the time of direct start.
2) Critical point $B$: $s = s_m$, the corresponding electromagnetic torque $T_m$ is the maximum torque that the motors can provide, which reflects the overload capacity of the motors. When the load torque exceeds the maximum torque of the motors, the motor would be forced to stop, and it is likely to cause an accident. It is important to note that although motors have certain overload ability, operating for a long time in the case of overload is forbidden, as the resultant high temperature will destroy the motors.

3) Rated working point *C*: in inherent characteristic curve, the speed, torque, current, and power corresponding to rated point *C* are all rated values. The slip $s_N$ corresponding to rated speed is called rated slip, ranging in 0.015–0.05. The rated torque on the mechanical characteristic curve refers to the rated electromagnetic torque, expressed as $T_N$.
4) Synchronization point *D*: synchronization point is also called ideal no-load point. On this point, $n = n_1 (s = 0)$, $T = 0$, $E_{2s} = 0$, $I_2 = 0$, $I_1 = I_m$. The motor is in ideal no-load condition and has no electromechanical energy conversion.

### 2.4.3 Man-Made Mechanical Characteristic of Asynchronous Motors

Man-made mechanical characteristic is the mechanical characteristic after changing either parameter of the motor artificially, such as changing the stator voltage $U_1$, stator frequency $f_1$, magnetic pole number $p$, resistance and reactance of stator loop, resistance and reactance of rotor loop, and so forth.

#### 2.4.3.1 Man-Made Characteristic of Reducing Stator Voltage

Man-made characteristic of reducing stator voltage is the characteristic on the condition of only reducing phase voltage retaining the other parameters as such with those of inherent characteristic. The man-made characteristic curves of different $U_1$ are shown in Figure 2.17.

It has following features:

1) Synchronous speed $n_1$ and critical slip $s_m$ keep invariant after reducing voltage, namely, the man-made characteristics of different $U_1$ all go through the ideal no-load point, and the maximum torques are obtained at the same speed.
2) After reducing voltage, the decrease of maximum torque $T'_m$ and starting torque $T'_Q$ are proportional to $U_1^2$.

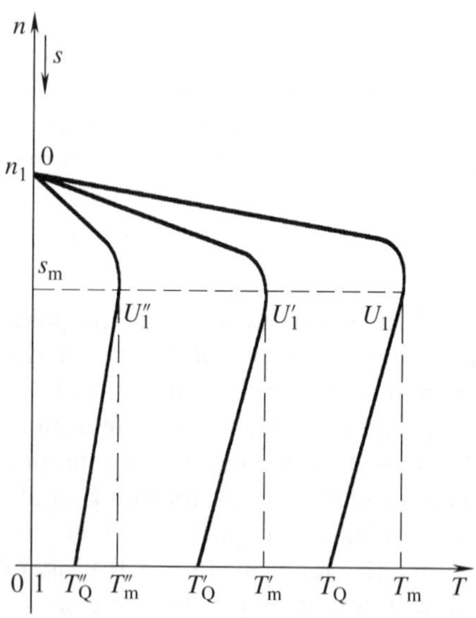


**Figure 2.17** Three-phase asynchronous motors' artificial characteristic of reducing voltage.

#### 2.4.3.2 Man-Made Characteristic of Connecting Symmetrical Three-Phase Resistances in the Rotor's Loop

For wound rotor asynchronous motors, man-made characteristic of connecting symmetrical three-phase resistances in the rotor's loop is the characteristics on the condition of only connecting symmetrical three-phase resistances in the rotor's loop but keeping the same parameters with those of inherent characteristics.

The practical expression of man-made characteristic of connecting symmetrical three-phase resistances in the rotor's loop is as follows:

$$T = \frac{2T_{\mathrm{m}}}{\dfrac{s}{s'_{\mathrm{m}}} + \dfrac{s'_{\mathrm{m}}}{s}}$$

$$\frac{s'_{\mathrm{m}}}{s_{\mathrm{m}}} \approx \frac{(r'_2 + R'_\Omega)/(x_{1\sigma} + x'_{2\sigma})}{r'_2/(x_{1\sigma} + x'_{2\sigma})} = \frac{r'_2 + R'_\Omega}{r'_2} = \frac{r_2 + R_\Omega}{r_2} = 1 + \frac{R_\Omega}{r_2}$$

Man-made characteristics curve of connecting different $R_\Omega$ is drawn out according to the equations above as depicted in Figure 2.18.

It has following features:

1) Synchronous speed $n_1$ of the motors keeps constant, so all the man-made characteristics curves of different $R_\Omega$ go through the ideal no-load point.
2) The critical slip is proportional to the rotor resistance, so $s'_{\mathrm{m}} > s_{\mathrm{m}}$, and it increases with the increase of $R_\Omega$, but $T_{\mathrm{m}}$ keeps unchanged.

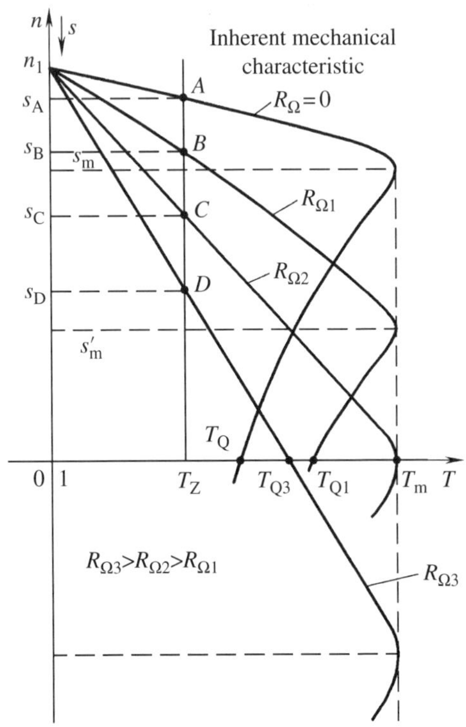


**Figure 2.18** Artificial characteristic of connecting symmetrical three-phase resistances in rotor's loop.

3) When $s'_m < 1$, $T_Q$ increases with the increase of $R_\Omega$; But when $s'_m > 1$, $T_Q$ reduces with the increase of $R_\Omega$ instead. Therefore, connecting appropriate resistance in series with rotor loop can improve the starting ability of the motors and improve the starting performance of asynchronous motors. For example, if we set $s'_m = 1$, in order to make $T_Q = T_m$, the resistance connected in series with the rotor of each phase according to (2.18) should be

$$R'_\Omega = \sqrt{r_1^2 + \left(x_{1\sigma} + x'_{2\sigma}\right)^2} - r'_2$$

#### 2.4.3.3 Man-Made Characteristic of Changing the Frequency of Stator Voltage

Because the synchronous speed of three-phase asynchronous motors is proportional to the frequency $f_1$ of the stator power source, changing $f_1$ can change the synchronous speed if the pole pairs of asynchronous motors keep constant, which can achieve the purpose of smooth speed control with a large speed regulation range. The rated frequency of the motor is the reference frequency, which is the dividing line in variable frequency speed control. At the same time, it can be divided into constant torque variable frequency speed control and constant power variable frequency speed control according to different control methods. According to the load characteristics, we can choose either constant torque variable frequency speed control or constant power variable frequency speed control to achieve the optimal effect.

1) $f_1 < f_N$, below reference frequency, constant torque variable frequency speed control of keeping $U_1/f_1$ = constant (main magnetic flux $\Phi_m$ keeps unchanged), at this time, the man-made mechanical characteristic curve of the motors is shown in Figure 2.19, and the straight line segment is parallel with that of inherent characteristics.

   The features of mechanical characteristics of variable frequency speed control under reference frequency are as follows:

   ① When frequency $f$ changes, the speed drop $\Delta n_m$ corresponding to the maximum torque keeps constant.
   ② Synchronous speed $n_1 = (60 f_1/p) \propto f$.
   ③ The maximum torque $T_m$ keeps constant.
   ④ Starting torque $T'_Q$ increases when frequency decreases according to the following equation:

$$T_Q \approx \frac{m_1 p r'_2}{8\pi^3 f_1\left(L_{1\sigma} + L'_{2\sigma}\right)} \propto \frac{1}{f}$$

   When the frequency is very low, the stator's $r_1$ cannot be ignored. If $U_1/f_1 = U'/f'_1$ = constant, when $f_1$ declines, the decreasing multiple of denominator is smaller than that of numerator, and $T_m$ decrease, the man-made characteristic is shown as the curve ③ is depicted in Figure 2.19. In order to avoid $T_m$ dropping too much, it usually needs to increase voltage $U'_1$ properly at low speed.

2) $f_1 > f_N$, over reference frequency, constant power variable frequency speed control of keeping $U_1/\sqrt{f_1}$ = constant.

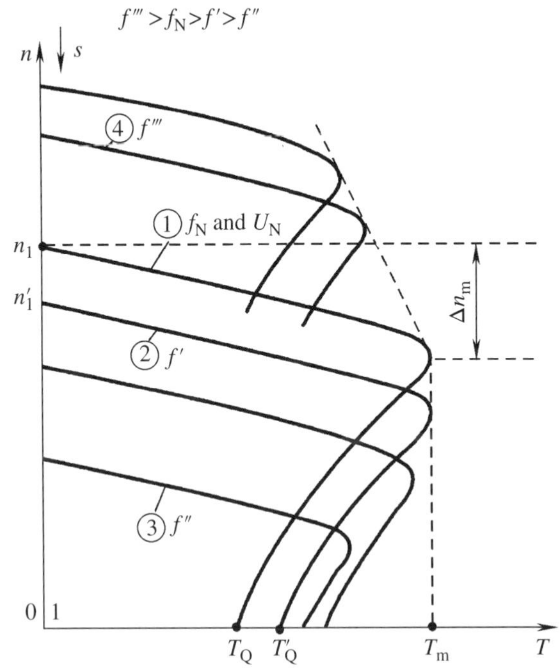


**Figure 2.19** Artificial characteristic of variable frequency speed control.

When controlling the motors speed by variable frequency above reference frequency, the stator frequency $f_1$ is greater than rated frequency $f_N$. To maintain $\Phi_m$ constant, stator voltage $U_1$ will be higher than rated voltage, which is not allowed. So when controlling the motor's speed by variable frequency above reference frequency, voltage $U_1$ should be maintained in rated value; in this way, $n$ increases with the increase of frequency $f_1$, $T_m$ and $T_Q$ reduce with the decrease of magnetic flux, which is approximate to constant power control. The man-made characteristic is shown as curve ④ depicted in Figure 2.19.

## 2.5 Start-up of Three-Phase Asynchronous Motors

### 2.5.1 Starting Requirements of Three-Phase Asynchronous Motors

The requirements for the starting performance of asynchronous motors mainly include following two aspects:

#### 2.5.1.1 In Order to Minimize the Impact on the Grid, the Starting Current Should be Small

If the motor is connected as the rated type with no impedance in series with the stator and rotor loop, and it is connected to the grid with rated voltage and frequency, we name this type of start as direct start, and the performance in the process of the motor rotating from stationary state to stable operation state is called inherent start characteristics. According to the starting equivalent circuit, the directly starting phase current of the stator is

$$I_{1Q} \approx I'_{2Q} = \frac{U_1}{\sqrt{\left(r_1 + r'_2\right)^2 + \left(x_{1\sigma} + x'_{2\sigma}\right)^2}} = \frac{U_1}{Z_k}$$

Because $U_1 = U_{N\varphi}$ is very high but $Z_k$ is very small, the direct starting current is very large, and the line current $I_Q$ is about 5–7 times of the rated current $I_N$. So the power supply transformer must provide a large starting current to make the output voltage of the power supply transformer decrease, which has an impact on the power grid. If the rated capacity of transformer is not large enough, the large starting current will make the transformer output voltage decline very soon to exceed the normal range, which will affect other loads supplied by the same transformer, and obviously, this is not allowed. So, when the rated capacity of power supply transformer is not big enough compared to the rated power of motors, the three-phase asynchronous motor is not allowed to directly start under the rated voltage, and it needs to take measures to reduce the starting current.

#### 2.5.1.2 The Starting Torque Must Be Large Enough to Speed Up the Starting Process and Shorten the Starting Time

In direct starting mode, because $\cos\varphi_2$ of start state is much smaller than that of rated operation, and the leakage impedance voltage drop $I_{1Q}Z_1$ is very large, the main magnetic flux $\Phi_m$ is much smaller than the rated value. According to the physical expression of torque, although starting current $I_2'$ is very big, the starting electromagnetic torque is relatively small, generally about 0.8 times of rated value, which has almost no impact in light load or no load condition, but may fail to start the motor in heavy load.

### 2.5.2 Conditions for Squirrel Cage Asynchronous Motors Starting Directly

Generally, for asynchronous motors with 380 V rated voltage and capacity less than 7.5 kW, they could start directly. For motors with capacity over 7.5 kW, the impact of overlarge direct starting current on power supply system must be considered, and generally verified by the following empirical formula:

$$\frac{I_Q}{I_N} \leq \frac{1}{4}\left(3 + \frac{S_H}{P_N}\right) \tag{2.25}$$

where $I_Q$, the stator line current when directly starting;
$I_N$, the rated current of the motors;
$S_H$, the total capacity of power supply (kVA);
$P_N$, the rated power of the motors (kW).

If formula (2.25) is satisfied, this motor is allowed to start directly; otherwise, measures must be taken to limit the starting current.

In order to correctly use the direct starting method, two indicators of the inherent starting features are provided in the production catalogue of cage asynchronous motors: the starting torque ratio $K_M$ and the starting current ratio $K_I$.

$$K_M = \frac{T_Q}{T_N}$$

$$K_I = \frac{I_Q}{I_N}$$

Here, $T_Q$ and $I_Q$ signify the starting torque and starting line current of stator during starting directly. For ordinary cage asynchronous motors, $K_M \approx 0.8–1.8$ and $K_I \approx 5–7$. So when choosing direct starting method, we must check whether $K_M$ and $K_I$ meet the system requirements.

In the industrial field application, besides the direct starting method, a large number of motors are suitable for other starting ways, which will be introduced in details in the subsequent chapters of the book.

## 2.6 Energy Efficiency Standards of Three-Phase Asynchronous Motors

The power consumption of motor system occupies a considerable proportion in the world and the whole industrial motors consume about 30–40% of the generated electricity. According to EU statistics, the motor system accounts for 42% of the total electricity consumption, and 69% of its industrial electricity consumption. So many countries, especially developed countries, have carried out the motor system energy conservation action. All countries in the world formulate the corresponding laws and regulations to improve the market share of efficient, even superefficient, motors. Efficient or superefficient motors have become mainstream in the international market, and the ordinary motors are phased out by the market.

In the United States, the standard market access system of efficient motors has been compulsorily carried out. At present, America has compulsorily popularized the adjustable Epact motors conforming to IE2 class and efficient motors NEMA conforming to IE3 class. The U.S. Department of Energy, environmental protection bureau, and the coalition of energy efficiency (CEE) have passed a resolution in March 2007 that the United States will impose superefficient motors standard (Energy Independence & Security Act 2007, EISAct) in 2011.

In Europe, EU-DGTE in 2006 decided to prohibit the production and circulation of Eff 3 class motors in future. On August 11, 2007, the European Union translated "Eco-Design of Energy-using Products" (EuP) formally into laws and regulations for EU member countries, including mandatory energy efficiency index for motors and other mechanical and electrical products. In 2009, the European Union ruled the minimum mandatory energy efficiency performance standards (MEPS) of the motors, replacing the IE1 motors by more efficient IE2, IE3 class efficient motors.

Australia also has issued a mandatory standard of energy conservation efficient motors. New Zealand also performs this standard. In Australia and New Zealand, the production and imports of motors must achieve or exceed the lowest efficiency index of the standards, and the index values are close to the Eff 1 of EU and Epact of the United States.

The International Electrotechnical Commission (IEC) established IEC60034-30, in 2008: 2008 motors efficiency standards, and it was taken into effect in 2009 around the world. In the standard, motors must reach or exceed the standard, defined by the IEC 60034-30, IE2 efficiency class before January 1, 2011 and reach or exceed IE3 efficiency class before January 1, 2015 (efficiency classes ruled by IEC60034-30 standard are divided into IE1, IE2, IE3, and IE4, where IE4 is the highest grade). The following will discuss the current efficiency standards of IEC three-phase asynchronous motors in details first.

### 2.6.1 Energy Efficiency Standards of IEC Three-Phase Asynchronous Motors

The IEC is a worldwide standardization organization made up of the electrical commissions (IEC National Committee) of all the countries. In October 2008, the IEC/TC2 rotating electrical motors technical commission released the IEC 60034-30 (1st edition)—efficiency classes of single-speed, three-phase, and cage-asynchronous motors (IE code).

This standard of IEC is to coordinate the efficiency classes of motors in the world. As mentioned earlier, there are many different standards of energy efficiency for cage asynchronous motors in use (NEMA, EPACT, CSA, CEMEP COPANT, the AS/NZS, JIS, GB, etc.), and the new standards are still coming up. For a global market, different standards of different countries are bound to increase the difficulty of design and manufacture for manufacturers, so the IEC released this standard to coordinate different requirements, and make motors' design of each efficiency class adapt to both 50 and 60 Hz power supply, which can reduce customer's confusion due to variable efficiency requirements.

#### 2.6.1.1 Standard Applicable Scope

This standard applies to efficiency classes for single-speed, three-phase, 50 and 60 Hz, cage-asynchronous motors that

- have a rated voltage up to 1000 V (the standard also applies to motors rated for two or more voltages);
- have a rated output power $P_N$ between 0.75 and 370 kW;
- have either 2, 4, or 6 poles;
- are rated on the basis of either duty type S1 (continuous duty) or S3 (intermittent periodic duty) with a rated cyclic duration factor of 80% or higher;
- are capable of operating directly connected to power grid;
- have rated operating conditions in accordance with IEC 60034-1, in Chapter 6.

#### 2.6.1.2 Class Standards

Efficiency Class is constituted by the letters "IE" (International Energy—efficiency Class, the abbreviation of "International Energy efficiency Class"), and the number which represents its grade, as shown in detail in Table 2.2.

**Table 2.2** IE-efficiency class.

| The digital code | A brief description | Definition |
|---|---|---|
| 1 | Standard efficiency | Motors in full load (rated output), rated efficiency is equal to or greater than the limit of the Table 2.3 |
| 2 | High efficiency | Motors in full load (rated output), rated efficiency is equal to or greater than the limit of the Table 2.4 |
| 3 | Super-high efficiency | Motors in full load (rated output), rated efficiency is equal to or greater than the limit of the Table 2.5 |
| 4 | Ultrasuper high efficiency | IE4 efficiency class will be merged in the standard text in the next version of this standard. The goal is to reduce 15% of loss on the basis of IE3. Technological progress or other motors are expected to reach to IE4. The scope of this standard will be modified according to needs |

**Table 2.3** IE1 standard efficiency stipulated limits (50 Hz) (%).

| | Pole number | | |
|---|---|---|---|
| $P_N$ (kW) | 2 | 4 | 6 |
| 0.75 | 72.1 | 72.1 | 70.0 |
| 1.1 | 75.0 | 75.0 | 72.9 |
| 1.5 | 77.2 | 77.2 | 75.2 |
| 2.2 | 79.7 | 79.7 | 77.7 |
| 3 | 81.5 | 81.5 | 79.7 |
| 4 | 83.1 | 83.1 | 81.4 |
| 5.5 | 84.7 | 84.7 | 83.1 |
| 7.5 | 86.0 | 86.0 | 84.7 |
| 11 | 87.6 | 87.6 | 86.4 |
| 15 | 88.7 | 88.7 | 87.7 |
| 18.5 | 89.3 | 89.3 | 88.6 |
| 22 | 89.9 | 89.9 | 89.2 |
| 30 | 90.7 | 90.7 | 90.2 |
| 37 | 91.2 | 91.2 | 90.8 |
| 45 | 91.7 | 91.7 | 91.4 |
| 55 | 92.1 | 92.1 | 91.9 |
| 75 | 92.7 | 92.7 | 92.6 |
| 90 | 93.0 | 93.0 | 92.9 |
| 110 | 93.3 | 93.3 | 93.3 |
| 132 | 93.5 | 93.5 | 93.5 |
| 160 | 93.8 | 93.8 | 93.8 |
| 200–370 | 94.0 | 94.0 | 94.0 |

#### 2.6.1.3 Interpolation Calculation

If the output power of the motors is not defined in the above tables, but in the range of 0.75–200 kW, its nominal efficiency limit can be determined by the following formula (suitable for 50 Hz).

$$\eta_N = A\times\left[\log_{10}(P_N)\right]^3 + B\times\left[\log_{10}(P_N)\right]^2 + C\times\log_{10}(P_N) + D$$

Here, $A$, $B$, $C$, $D$ are the interpolation coefficients (in Table 2.6), and the unit of $P_N$ is kW. The efficiency value should be described as: xx.x%. In addition, the formula and the interpolation coefficients are fitted by the mathematical methods, just to facilitate the interpolation calculation but without physical significances.

**Table 2.4** IE2 high-efficiency stipulated limit (50 Hz) (%).

| | Pole number | | |
|---|---|---|---|
| $P_N$ (kW) | 2 | 4 | 6 |
| 0.75 | 77.4 | 79.6 | 75.9 |
| 1.1 | 79.6 | 81.4 | 78.1 |
| 1.5 | 81.3 | 82.8 | 79.8 |
| 2.2 | 83.2 | 84.3 | 81.8 |
| 3 | 84.6 | 85.5 | 83.3 |
| 4 | 85.8 | 86.6 | 84.6 |
| 5.5 | 87.0 | 87.7 | 86.0 |
| 7.5 | 88.1 | 88.7 | 87.2 |
| 11 | 89.4 | 89.8 | 88.7 |
| 15 | 90.3 | 90.6 | 89.7 |
| 18.5 | 90.9 | 91.2 | 90.4 |
| 22 | 91.3 | 91.6 | 90.9 |
| 30 | 92.0 | 92.3 | 91.7 |
| 37 | 92.5 | 92.7 | 92.2 |
| 45 | 92.9 | 93.1 | 92.7 |
| 55 | 93.2 | 93.5 | 93.1 |
| 75 | 93.8 | 94.0 | 93.7 |
| 90 | 94.1 | 94.2 | 94.0 |
| 110 | 94.3 | 94.5 | 94.3 |
| 132 | 94.6 | 94.7 | 94.6 |
| 160 | 94.8 | 94.9 | 94.8 |
| 200–370 | 95.0 | 95.1 | 95.0 |

### 2.6.2 Energy Efficiency Standards of Three-Phase Asynchronous Motors in the United States and EU

Here are the energy efficiency standards of the three-phase asynchronous motors in the United States and EU for reference.

1) Energy efficiency standards of motors in the United States (Table 2.7).
2) Energy efficiency standards of motors in EU (Table 2.8).

### 2.6.3 Energy Efficiency Standards of Three-Phase Asynchronous Motors in China

At present, the installed capacity of all kinds of motors has been more than 500 million kW in China, among which asynchronous motors account for about 90%, medium and small motors account for about 80%, and the capacity of mechanical motors of

**Table 2.5** IE3 ultra high-efficiency stipulated limits (50 Hz) (%).

| | Pole number | | |
|---|---|---|---|
| $P_N$ (kW) | 2 | 4 | 6 |
| 0.75 | 80.7 | 82.5 | 78.9 |
| 1.1 | 82.7 | 84.1 | 81.0 |
| 1.5 | 84.2 | 85.3 | 82.5 |
| 2.2 | 85.9 | 86.7 | 84.3 |
| 3 | 87.1 | 87.7 | 85.6 |
| 4 | 88.1 | 88.6 | 86.8 |
| 5.5 | 89.2 | 89.6 | 88.0 |
| 7.5 | 90.1 | 90.4 | 89.1 |
| 11 | 91.2 | 91.4 | 90.3 |
| 15 | 91.9 | 92.1 | 91.2 |
| 18.5 | 92.4 | 92.6 | 91.7 |
| 22 | 92.7 | 93.0 | 92.2 |
| 30 | 93.3 | 93.6 | 92.9 |
| 37 | 93.7 | 93.9 | 93.3 |
| 45 | 94.0 | 94.2 | 93.7 |
| 55 | 94.3 | 94.6 | 94.1 |
| 75 | 94.7 | 95.0 | 94.6 |
| 90 | 95.0 | 95.2 | 94.9 |
| 110 | 95.2 | 95.4 | 95.1 |
| 132 | 95.4 | 95.6 | 95.4 |
| 160 | 95.6 | 95.8 | 95.6 |
| 200–370 | 95.8 | 96.0 | 95.8 |

**Table 2.6** Interpolation coefficients.

| | | 0.75–200 kW | | |
|---|---|---|---|---|
| IE-code | Coefficients | 2 Poles | 4 Poles | 6 Poles |
| IE1 | *A* | 0.5234 | 0.5234 | 0.0786 |
| | *B* | −5.0499 | −5.0499 | −3.5838 |
| | *C* | 17.4180 | 17.4180 | 17.2918 |
| | *D* | 74.3171 | 74.3171 | 72.2383 |
| IE2 | *A* | 0.4211 | 0.2252 | 0.0148 |
| | *B* | −3.9030 | −2.8026 | −2.4978 |
| | *C* | 13.7008 | 11.3920 | 13.2470 |
| | *D* | 79.0554 | 80.9783 | 77.5603 |
| IE3 | *A* | 0.3569 | 0.2162 | 0.1620 |
| | *B* | −3.3076 | −2.5259 | −2.5759 |
| | *C* | 11.6108 | 9.9665 | 11.3839 |
| | *D* | 82.2503 | 83.7712 | 81.1773 |

**Table 2.7** American motors (EPACT) efficiency value (%).

| | 2 Poles | | 4 Poles | | 6 Poles | |
|---|---|---|---|---|---|---|
| Power (hp) | Rated value | Minimum value | Rated value | Minimum value | Rated value | Minimum value |
| 1.0 | 75.5 | 74.0 | 82.5 | 81.5 | 80.0 | 78.5 |
| 1.5 | 82.5 | 81.5 | 84.0 | 82.5 | 85.5 | 84 |
| 2.0 | 84.0 | 82.5 | 84.0 | 82.5 | 86.5 | 85.5 |
| 3.0 | 85.5 | 84.0 | 87.5 | 86.5 | 87.5 | 86.5 |
| 5.0 | 87.5 | 86.5 | 87.5 | 86.5 | 87.5 | 86.5 |
| 7.5 | 88.5 | 87.5 | 89.5 | 88.5 | 89.5 | 88.5 |
| 10.0 | 89.5 | 88.5 | 89.5 | 88.5 | 89.5 | 88.5 |
| 15.0 | 90.2 | 89.5 | 91.0 | 90.2 | 90.2 | 89.5 |
| 20.0 | 90.2 | 89.5 | 91.0 | 90.2 | 90.2 | 89.5 |
| 25.0 | 91.0 | 90.2 | 92.4 | 91.7 | 91.7 | 91.0 |
| 30.0 | 91.0 | 90.2 | 92.4 | 91.7 | 91.7 | 91.0 |
| 40.0 | 91.7 | 91.0 | 93.0 | 92.4 | 93.0 | 92.4 |
| 50.0 | 92.4 | 91.7 | 93.0 | 92.4 | 93.0 | 92.4 |
| 60.0 | 93.0 | 92.4 | 93.6 | 93.0 | 93.6 | 93.0 |
| 75.0 | 93.0 | 92.4 | 94.1 | 93.6 | 93.6 | 93.0 |
| 100.0 | 93.6 | 93.0 | 94.5 | 94.1 | 94.1 | 93.6 |
| 125.0 | 94.5 | 94.1 | 94.5 | 94.1 | 94.5 | 94.1 |
| 150.0 | 94.5 | 94.1 | 95.0 | 94.5 | 95.0 | 94.5 |
| 200.0 | 95.0 | 94.5 | 95.0 | 94.5 | 95.0 | 94.5 |

drag-fans, pumps, and compressors is about 200 million kW. According to the statistics, at present, the operating efficiency of all kinds of motors in China is 3–5% lower than the average efficiency level of developed countries, and the efficiency of fans and pumps in China is 2–3% lower than that in developed countries. Therefore, in the past dozen years, a lot of work has been carried out in motor energy-conversation in China. About 63 kinds of high energy-consuming motors have been forbidden and 24 kinds of energy-saving motors have been promoted.

At the same time, in order to promote the continuous development of electric motor energy conservation, China had formulated the GB18613-2002 “the minimum allowable values of energy efficiency and evaluating values of energy saving for small and medium three-phase asynchronous motors” in 2002, which restricts the production and use of inefficient motors. In December 2006, the standard was revised published as GB18613-2006 “the minimum allowable values of energy efficiency and the energy efficiency grades for small and medium three-phase asynchronous motors.” And it stipulates that from January 1, 2010, the minimum allowable energy efficiency values of energy-saving motors would be put into practice.

**Table 2.8** EU motors (CEMEP) efficiency value (%).

| | 2 Poles | | | 4 Poles | | |
|---|---|---|---|---|---|---|
| **Power (kW)** | **Eff 3** | **Eff 2** | **Eff 1** | **Eff 3** | **Eff 2** | **Eff 1** |
| 1.1 | <76.2 | ≥76.2 | ≥82.8 | <76.2 | ≥76.2 | ≥83.8 |
| 1.5 | <78.5 | ≥78.5 | ≥84.1 | <78.5 | ≥78.5 | ≥85.0 |
| 2.2 | <81.0 | ≥81.0 | ≥85.6 | <81.0 | ≥81.0 | ≥86.4 |
| 3 | <82.6 | ≥82.6 | ≥86.7 | <82.6 | ≥82.6 | ≥87.4 |
| 4 | <84.2 | ≥84.2 | ≥87.6 | <84.2 | ≥84.2 | ≥88.3 |
| 5.5 | <85.7 | ≥85.7 | ≥88.6 | <85.7 | ≥85.7 | ≥89.2 |
| 7.5 | <87.0 | ≥87.0 | ≥89.5 | <87.0 | ≥87.0 | ≥90.1 |
| 11 | <88.4 | ≥88.4 | ≥90.5 | <88.4 | ≥88.4 | ≥91.0 |
| 15 | <89.4 | ≥89.4 | ≥91.3 | <89.4 | ≥89.4 | ≥91.8 |
| 18.5 | <90.0 | ≥90.0 | ≥91.8 | <90.0 | ≥90.0 | ≥92.2 |
| 22 | <90.5 | ≥90.5 | ≥92.2 | <90.5 | ≥90.5 | ≥92.6 |
| 30 | <91.4 | ≥91.4 | ≥92.9 | <91.4 | ≥91.4 | ≥93.2 |
| 37 | <92.0 | ≥92.0 | ≥93.3 | <92.0 | ≥92.0 | ≥93.6 |
| 45 | <92.5 | ≥92.5 | ≥93.7 | <92.5 | ≥92.5 | ≥93.9 |
| 55 | <93.0 | ≥93.0 | ≥94.0 | <93.0 | ≥93.0 | ≥94.2 |
| 75 | <93.6 | ≥93.6 | ≥94.6 | <93.6 | ≥93.6 | ≥94.7 |
| 90 | <93.9 | ≥93.9 | ≥95.0 | <93.9 | ≥93.9 | ≥95.0 |

The GB18613-2006 "the minimum allowable values of energy efficiency and the energy efficiency grades for small and medium three-phase asynchronous motors" stipulates energy efficiency grades, minimum energy efficiency allowable values, target minimum energy efficiency allowable values, evaluating values of energy saving, and experimental methods of the small and medium three-phase asynchronous motors.

The efficiency of the three-phase induction motors is divided into three grades, in which grade 1 has the highest energy efficiency. The minimum energy efficiency allowable values of motors in each grade in conditions of rated power output and 75% rated output power should not be less than the provisions in Table 2.9.

The minimum energy efficiency allowable value is defined as the minimum efficiency allowable value under standard test conditions. The minimum energy efficiency allowable values of motors in conditions of rated power output and 75% rated output power should not be less than the provision of grade 3 in Table 2.9.

The definition of target minimum energy efficiency allowable value of motors is the allowable minimum efficiency of motors after implementing this standard of a certain number of years, which will replace minimum energy efficiency allowable value once used. Also, the target minimum energy efficiency allowable value of motors in conditions of rated power output and 75% rated output power should not be less than the provisions of grade 2 in Table 2.9.

**Table 2.9** Efficiency standards for three-phase asynchronous motors in China (%).

| | 1 Grade | | | 2 Grade | | | 3 Grade | | |
|---|---|---|---|---|---|---|---|---|---|
| **Power (kW)** | **2 Poles** | **4 Poles** | **6 Poles** | **2 Poles** | **4 Poles** | **6 Poles** | **2 Poles** | **4 Poles** | **6 Poles** |
| 0.55 | — | — | — | — | 80.7 | 75.4 | — | 71.0 | 65.0 |
| 0.75 | — | — | — | 77.5 | 82.3 | 77.7 | 75.0 | 73.0 | 69.0 |
| 1.1 | — | — | — | 82.8 | 83.8 | 79.9 | 76.2 | 76.2 | 72.0 |
| 1.5 | — | — | — | 84.1 | 85.0 | 81.5 | 78.5 | 78.5 | 76.0 |
| 2.2 | — | — | — | 85.6 | 86.4 | 83.4 | 81.0 | 81.0 | 79.0 |
| 3 | — | — | 86.9 | 86.7 | 87.4 | 84.9 | 82.6 | 82.6 | 81.0 |
| 4 | 89.3 | 89.9 | 87.9 | 87.6 | 88.3 | 86.1 | 84.2 | 84.2 | 82.0 |
| 5.5 | 90.1 | 90.7 | 89.1 | 88.6 | 89.2 | 87.4 | 85.7 | 85.7 | 84.0 |
| 7.5 | 90.9 | 91.5 | 90.6 | 89.5 | 90.1 | 89.0 | 87.0 | 87.0 | 86.0 |
| 11 | 91.9 | 92.2 | 91.4 | 90.5 | 91.0 | 90.0 | 88.4 | 88.4 | 87.5 |
| 15 | 92.5 | 92.9 | 92.3 | 91.3 | 91.8 | 91.0 | 89.4 | 89.4 | 89.0 |
| 18.5 | 92.9 | 93.3 | 92.7 | 91.8 | 92.2 | 91.5 | 90.0 | 90.0 | 90.0 |
| 22 | 93.3 | 93.6 | 93.1 | 92.2 | 92.6 | 92.0 | 90.5 | 90.5 | 90.0 |
| 30 | 93.9 | 94.2 | 93.6 | 92.2 | 93.2 | 92.5 | 91.4 | 91.4 | 91.5 |
| 37 | 94.2 | 94.5 | 94.0 | 93.3 | 93.6 | 93.0 | 92.0 | 92.0 | 92.0 |
| 45 | 94.6 | 94.8 | 94.4 | 93.7 | 93.9 | 93.5 | 92.5 | 92.5 | 92.5 |
| 55 | 94.9 | 95.0 | 94.7 | 94.0 | 94.2 | 93.8 | 93.0 | 93.0 | 92.8 |
| 75 | 95.4 | 95.5 | 95.0 | 94.6 | 94.7 | 94.2 | 93.6 | 93.6 | 93.5 |
| 90 | 95.5 | 95.7 | 95.2 | 95.0 | 95.0 | 94.5 | 93.9 | 93.9 | 93.8 |
| 110 | 95.8 | 96.1 | 95.7 | 95.0 | 95.4 | 95.0 | 94.0 | 94.5 | 94.0 |
| 132 | 96.1 | 96.1 | 95.7 | 95.4 | 95.4 | 95.0 | 94.5 | 94.8 | 94.2 |
| 160 | 96.1 | 96.1 | 95.7 | 95.4 | 95.4 | 95.0 | 94.6 | 94.9 | 94.5 |
| 200 | 96.1 | 96.1 | 95.7 | 95.4 | 95.4 | 95.0 | 94.8 | 94.9 | 94.5 |
| 250 | 96.1 | 96.1 | 95.7 | 95.8 | 95.8 | 95.0 | 95.2 | 95.2 | 94.5 |
| 315 | 96.1 | 96.1 | — | 95.8 | 95.8 | — | 95.4 | 95.2 | — |

The definition of motor's evaluating values of energy saving is the minimum guarantee value of the motor's efficiency that satisfies the requirements of energy-saving certification in the standard test conditions. The motor's evaluating values of energy saving in conditions of rated power output and 75% rated output power should not be less than the provisions of grade 2 in Table 2.9.

The energy efficiency ratings of three-phase asynchronous motors are very important—they are the quality inspection standards for products, the references for users to choose motors, and foundations for energy-saving analysis of motors.

## 2.7 Mainstream Products of Three-Phase Asynchronous Motors

The Y-series motor takes about 90% of the market share in China, and the average efficiency of Y-series motor is about 87.3%. Average efficiency of American efficient machines is 90.3%, which is the market access level that was ruled by the US energy policy. The average efficiency of American superefficient motors is 91.7%. The efficiency of Y series motor of China is about 3% lower than that of efficient motors of the United States and other developed countries, and is about 4.4% lower than that of American superefficient motors. About 300 series and 1500 kinds of motors are being produced in China by about 300 known enterprises. The main products include the Y series (average efficiency of 87.72%), Y2 series (average efficiency of 87.73%), Y3 series (average efficiency of 87.75%) of common-efficient series, and Y2-E, YX3 series (average efficiency of 90%, equivalent to the EU's Eff 1 level) of high-efficient series. Y series motors which are widely used in China, were designed in 1980s, Y2 series were designed in 1990s, and Y3 series are basically up to the international advanced level of the same kind motors in early twenty-first century which can achieve the IE1 grade (the minimum grade of energy efficiency ruled by IEC 60034-30 standard, equivalent to energy efficiency level of the EU's Eff 2 standards). At present, YX3 series products can basically achieve efficiency level of IE2 (equivalent to the EU's Eff 1 level) and are widely promoted. According to the statistics, the current market share of high-efficiency motors is only about 1%, which means more than 98% of the motors used are common-efficient motors (3% lower than high-efficiency motors of the United States, 5% lower than the efficiency of superefficient motors). But in the United States and Canada, the market share of high efficiency motors and super high efficiency motors are about 54 and 16%, respectively. Therefore, market for high efficiency motor in China is very promising.

Although currently more than 40 domestic enterprises can produce high-efficiency motors, the market share of high-efficiency motors is still small, only accounts for 10.4%, and quite a number of high-efficiency motors produced in China are to meet overseas demand.

### 2.7.1 Brief Introduction of Existing Products of Three-Phase Asynchronous Motors

At present, main representative products of the asynchronous motors in China are:

1) Y series asynchronous motors, whose capacity is usually from 0.55 to 200 kW, have B insulation grade and IP44 protection grade, reaching to the international electrotechnical commission (IEC) standards and the international level in the late 1970s. The market share is around 10% in 2010.
2) The YX series high-efficiency motors, whose capacity is from 1.5 to 90 kW, have 2, 4, and 6 poles. The efficiency of YX series motors is average 3% higher than that of Y series, and is close to the international advanced level. YX series is suitable for conditions of single-way rotation, 3000 h and beyond annual operation time and 50% beyond load factor. The market share is around 30% in 2010.

### 2.7.2 Characteristics of Main Series of Three Phase Asynchronous Motors

As mentioned earlier, at present, the widely used products of three-phase asynchronous motors are Y, Y2, Y3 series in China. Though widely used, Y series motors still have a certain gap between foreign advanced level due to the defections in vibration, noise, volume, weight, reliability, protection grade, locked-rotor torque, and appearance quality. Y2 series motors, designed in 1993, have the highest efficiency and the best energy-saving effects with 50–75% of rated load and are suitable for conditions of 3000 and beyond operation hours. It has 1.5–2% higher efficiency than Y series, and the same average power factors with Y series, and its average locked-rotor torque is about 70% of the Y series. Besides, Y2 series could save about 13% of effective materials, has 10% less weight and less noise compared with Y series. The technical level of it could reach the foreign advanced level in the 1990s, not below the ILA5.6 series level of Siemens, and its power rating and installation dimensions are in accordance with IEC standard. Y3 series motors are the first products under the Chinese industrial policies "replacing hot rolled silicon steel with cold rolled silicon steel" that the whole series products are made of cold-rolled silicon steel sheet as permeability material, with a totally enclosed system, fan cooled outside squirrel cage structure, and have the characteristics of low noise, high torque, good starting performance, strong instantaneous overload capacity, compact structure, convenience of operation, and maintenance. Also, the efficiency index reaches to the European Eff 2 standards. YX3 series high-efficiency three-phase asynchronous motors were designed and developed on the basis of Y3 series motors by the Shanghai electric appliance science research institute (group) Co., Ltd., and a dozen representative companies of small and medium motors industry, its efficiency index fully meet the requirements of the European energy efficiency standard Eff l, and the noise level of YX3 series motors has increased greatly than the basic series (Y3 series), reducing the gap between foreign advanced countries. The main performance index has reached the international advanced level of similar products in twenty-first century. The main technical parameters and features of these several series products are shown in Tables 2.10 and 2.11.

### 2.7.3 Main Technical Data of Y2 Series Three-Phase Asynchronous Motors

To choose an appropriate kind of motor is the first thing in energy saving of three-phase asynchronous motors. Besides understanding the efficiency class of three-phase asynchronous motors and the efficiency index of every series (Y Y2 YX3), more understandings of three-phase asynchronous motor products are needed to choose the appropriate one. For example, if we select a 30 kW/6-pole motor preliminarily according to the load, we also need to consider rated torque according to the rated speed, the starting torque, minimum torque, starting current, input power, and so forth. Tables 2.12, 2.13, 2.14, and 2.15 are the technical data of Y2 series three-phase asynchronous motors with 2, 4, 6, and 8 poles, respectively for reference.

When choosing and using motors, in addition to the above technical data, protection grade, installation method, and installation size should also be considered. About these data, please refer to series technology condition of three-phase asynchronous motors in Chinese national standard and technical data of manufacturers.

**Table 2.10** Main technical parameters and features of Y, Y2, and Y3 series.

| Series products | Y series | Y2 series | Y3 series |
|---|---|---|---|
| Power (kW) | 0.55–200 | 0.12–315 | 0.12–315 |
| Center height of engine base | 80–315 mm, a total of 12 stands | 63–355 mm, a total of 15 stands | 63–355 mm, a total of 15 stands |
| Basic poles | 2, 4, 6, 8, 10 | 2, 4, 6, 8, 10 | 2, 4, 6, 8, 10 |
| Product specification number | A total of 68 specifications | A total of 117 specifications | A total of 117 specifications |
| Rated voltage | 380 V | 380 V | 380 V |
| Rated frequency | 50 Hz | 50 Hz | 50 Hz |
| Insulation grade | B grade | F grade | F grade |
| Protection form | IP44 | IP54 | IP55 |
| Structure shape | Radial distribution of stand radiating rib, cast iron base, junction box on the side | Vertical and parallel distribution of stand radiating rib; cast iron stand, H63–112 aluminum alloy frame; junction box on the side or top; H180–280 bearing lubrication with non-stop oiling structure | Vertical and parallel distribution of stand radiating rib; cast iron stand, H63–112 aluminum alloy frame; junction box on the side or top; H180–280 bearing lubrication with non-stop oiling structure |
| Noise level | Assessment of no-load noise | Assessment of no-load and load noise | Assessment of no-load and load noise |
| Silicon steel sheet material | Hot rolled silicon steel sheet | Hot rolled silicon steel sheet | Cold rolled silicon steel sheet |
| Standard number | JB/T10391-2002 | JB/T8680.1-1998 | JB/T10447-2004 |

## 2.8 Main Subseries Three-Phase Asynchronous Motors in China

Small and medium type three-phase asynchronous motors are the most widely used and the most productive among all kinds of motors. At present, the small and medium type three-phase asynchronous motors have grown to 300 series, more than 1500 varieties and 40,000 specifications in China. These series and varieties are all subseries of Y series according to the special requirements of all walks of life. They mainly include electric subseries, structure subseries, special environment subseries, and dedicated subseries. At present, there are 18 series of commonly used asynchronous motors, and their categories, characteristics, and applications are shown in Table 2.16. With the continuous development of science and technology, the three-phase asynchronous motors products series are also constantly changing. For example, because of the promotion of

**Table 2.11** The efficiency of Y, Y2, and YX3 series three-phase asynchronous motors.

| Rated power (kW) | Efficiency | | | | | | | | |
|---|---|---|---|---|---|---|---|---|---|
| | YX3 series | | | Y2 series | | | Y series | | |
| | 2 Poles | 4 Poles | 6 Poles | 2 Poles | 4 Poles | 6 Poles | 2 Poles | 4 Poles | 6 Poles |
| 0.55 | — | 80.7 | — | 73.0 | 71.0 | 65.0 | — | 73.0 | — |
| 0.75 | 77.5 | 82.3 | 77.7 | 75.0 | 73.0 | 69.0 | 75.0 | 74.5 | 72.5 |
| 1.1 | 82.8 | 83.8 | 79.9 | 77.0 | 75.0 | 72.0 | 77.0 | 78.0 | 73.5 |
| 1.5 | 84.1 | 85.0 | 81.5 | 79.0 | 78.0 | 76.0 | 78.0 | 79.0 | 77.5 |
| 2.2 | 85.6 | 86.4 | 83.4 | 81.0 | 80.0 | 79.0 | 80.5 | 81.0 | 80.5 |
| 3 | 86.7 | 87.4 | 84.9 | 83.0 | 82.0 | 81.0 | 82.0 | 82.5 | 83.0 |
| 4 | 87.6 | 88.3 | 86.1 | 85.0 | 84.0 | 82.0 | 85.5 | 84.5 | 84.0 |
| 5.5 | 88.6 | 89.2 | 87.4 | 86.0 | 85.0 | 84.0 | 85.5 | 85.5 | 85.3 |
| 7.5 | 89.5 | 90.1 | 89.0 | 87.0 | 87.0 | 86.0 | 86.2 | 87.0 | 86.0 |
| 11 | 90.5 | 91.0 | 90.0 | 88.0 | 88.0 | 87.5 | 87.2 | 88.0 | 87.0 |
| 15 | 91.3 | 91.8 | 91.0 | 89.0 | 89.0 | 89.0 | 88.2 | 88.5 | 89.5 |
| 18.5 | 91.8 | 92.2 | 91.5 | 90.0 | 90.5 | 90.0 | 89.0 | 91.0 | 89.8 |
| 22 | 92.2 | 92.6 | 92.0 | 90.0 | 91.0 | 90.0 | 89.0 | 91.5 | 90.2 |
| 30 | 92.9 | 93.2 | 92.5 | 91.2 | 92.0 | 91.5 | 90.0 | 92.2 | 90.2 |
| 37 | 93.3 | 93.6 | 93.0 | 92.0 | 92.5 | 92.0 | 90.5 | 91.8 | 90.8 |
| 45 | 93.7 | 93.9 | 93.5 | 92.3 | 92.8 | 92.5 | 91.5 | 92.3 | 92.0 |
| 55 | 94.0 | 94.2 | 93.8 | 92.5 | 93.0 | 92.8 | 91.5 | 92.6 | 92.0 |
| 75 | 94.6 | 94.7 | 94.2 | 93.0 | 93.8 | 93.5 | 92.0 | 92.7 | 92.8 |
| 90 | 95.0 | 95.0 | 94.5 | 93.8 | 94.2 | 93.8 | 92.5 | 93.5 | 93.2 |
| 110 | 95.0 | 95.4 | 95.0 | 94.0 | 94.5 | 94.0 | 92.5 | 93.5 | 93.5 |
| 132 | 95.4 | 95.4 | 95.0 | 94.5 | 94.8 | 94.2 | 93.0 | 94.0 | 93.8 |
| 160 | 95.4 | 95.4 | 95.0 | 94.6 | 94.9 | 94.5 | 93.5 | 94.5 | — |
| 200 | 95.4 | 95.4 | 95.0 | 94.8 | 95.0 | 94.7 | 93.5 | 94.5 | — |
| 250 | 95.8 | 95.8 | 95.0 | 95.3 | 95.3 | 94.9 | — | — | — |
| 315 | 95.8 | 95.8 | — | 95.6 | 95.6 | — | — | — | — |

variable frequency speed control technology, the electromagnetic speed control motors in old series have been replaced; with the development of the permanent magnet servo motors and control technology, electrical valve asynchronous motors have been replaced; with the application of permanent magnet synchronous motors variable frequency speed control technology, elevator asynchronous motors will be replaced gradually; with the promotion and application of variable frequency speed control technology in all walks of life, the variable frequency speed control (YVF2) motors series are added.

**Table 2.12** Technical data of Y2 series 2-pole three-phase asynchronous motors.

| Type | Rated power (kW) | Rated load | | | | Stall current/ rated current | Stall torque/ rated torque | Maximum torque/ rated torque | Minimum torque/ rated torque |
|---|---|---|---|---|---|---|---|---|---|
| | | Speed (r/min) | Current (A) | Efficiency (%) | Power factor | | | | |
| Y2-801-2 | 0.75 | 2830 | 1.78 | 75.0 | 0.83 | 6.1 | 2.2 | 2.3 | 1.5 |
| Y2-802-2 | 1.1 | 2830 | 2.49 | 77.0 | 0.84 | 7.0 | 2.2 | 2.3 | 1.5 |
| Y2-90S-2 | 1.5 | 2840 | 3.34 | 79.0 | 0.84 | 7.0 | 2.2 | 2.3 | 1.5 |
| Y2-90L-2 | 2.2 | 2840 | 4.69 | 81.0 | 0.85 | 7.0 | 2.2 | 2.3 | 1.4 |
| Y2-100L-2 | 3 | 2860 | 6.14 | 83.0 | 0.87 | 7.5 | 2.2 | 2.3 | 1.4 |
| Y2-112M-2 | 4 | 2880 | 7.83 | 85.0 | 0.88 | 7.5 | 2.2 | 2.3 | 1.4 |
| Y2-132S1_2 | 5.5 | 2900 | 10.7 | 86.0 | 0.88 | 7.5 | 2.2 | 2.3 | 1.2 |
| Y2-132S2-2 | 7.5 | 2900 | 14.2 | 87.0 | 0.88 | 7.5 | 2.2 | 2.3 | 1.2 |
| Y2-160M1-2 | 11 | 2930 | 20.9 | 88.0 | 0.89 | 7.5 | 2.2 | 2.3 | 1.2 |
| Y2-160M2-2 | 15 | 2930 | 27.9 | 89.0 | 0.89 | 7.5 | 2.2 | 2.3 | 1.2 |
| Y2-160L-2 | 18.5 | 2930 | 33.9 | 90.0 | 0.90 | 7.5 | 2.2 | 2.3 | 1.1 |
| Y2-180M-2 | 22 | 2940 | 40.5 | 90.0 | 0.90 | 7.5 | 2.0 | 2.3 | 1.1 |
| Y2-200L1-2 | 30 | 2950 | 54.8 | 91.2 | 0.90 | 7.5 | 2.0 | 2.3 | 1.1 |
| Y2-200L2-2 | 37 | 2950 | 66.6 | 92.0 | 090 | 7.5 | 2.0 | 2.3 | 1.1 |
| Y2-225M-2 | 45 | 2970 | 81.0 | 92.3 | 0.90 | 7.5 | 2.0 | 2.3 | 1.0 |
| Y2-250M-2 | 55 | 2970 | 99.6 | 92.5 | 0.90 | 7.5 | 2.0 | 2.3 | 1.0 |
| Y2-280S-2 | 75 | 2970 | 133.3 | 93.0 | 0.90 | 7.5 | 2.0 | 2.3 | 0.9 |
| Y2-280M-2 | 90 | 2970 | 158.2 | 93.8 | 0.91 | 7.5 | 2.0 | 2.3 | 0.9 |
| Y2-315S-2 | 110 | 2980 | 195.1 | 94.0 | 0.91 | 7.1 | 1.8 | 2.2 | 0.9 |
| Y2-315M-2 | 132 | 2980 | 231.6 | 94.5 | 0.91 | 7.1 | 1.8 | 2.2 | 0.9 |
| Y2-315L1-2 | 160 | 2980 | 279.6 | 94.6 | 0.92 | 7.1 | 1.8 | 2.2 | 0.9 |
| Y2-315L2-2 | 200 | 2980 | 347.7 | 94.8 | 0.92 | 7.1 | 1.8 | 2.2 | 0.8 |
| Y2-355M-2 | 250 | 2980 | 429.4 | 95.3 | 0.92 | 7.1 | 1.6 | 2.2 | 0.8 |
| Y2-355L-2 | 315 | 2980 | 538.8 | 95.6 | 0.92 | 7.1 | 1.6 | 2.2 | 0.8 |

**Table 2.13** Technical data of Y2 series 4-pole three-phase asynchronous motors.

| Type | Rated power (kW) | Rated load | | | | Stall current/ rated current | Stall torque/ rated torque | Maximum torque/ rated torque | Minimum torque/ rated torque |
|---|---|---|---|---|---|---|---|---|---|
| | | Speed (r/min) | Current (A) | Efficiency (%) | Power factor | | | | |
| Y2-802-4 | 0.75 | 1400 | 1.99 | 73.0 | 0.76 | 6.0 | 2.3 | 2.3 | 1.6 |
| Y2-90S-4 | 1.1 | 1400 | 2.80 | 75.0 | 0.77 | 6.0 | 2.3 | 2.3 | 1.6 |
| Y2-90L-4 | 1.5 | 1400 | 3.65 | 78.0 | 0.79 | 6.0 | 2.3 | 2.3 | 1.5 |
| Y2-100L1-4 | 2.2 | 1420 | 5.05 | 80.0 | 0.81 | 7.0 | 2.3 | 2.3 | 1.5 |
| Y2-100L2-4 | 3 | 1430 | 6.64 | 82.0 | 0.82 | 7.0 | 2.3 | 2.3 | 1.5 |
| Y2-112M-4 | 4 | 1440 | 8.62 | 84.0 | 0.82 | 7.0 | 2.3 | 2.3 | 1.4 |
| Y2-132S-4 | 5.5 | 1440 | 11.5 | 85.0 | 0.83 | 7.0 | 2.3 | 2.3 | 1.4 |
| Y2-132M-4 | 7.5 | 1440 | 15.3 | 87.0 | 0.84 | 7.0 | 2.3 | 2.3 | 1.4 |
| Y2-160M-4 | 11 | 1460 | 22.2 | 88.0 | 0.84 | 7.0 | 2.2 | 2.3 | 1.4 |
| Y2-160L-4 | 15 | 1460 | 29.8 | 89.0 | 0.85 | 7.5 | 2.2 | 2.3 | 1.4 |
| Y2-180M-4 | 18.5 | 1470 | 36.1 | 90.5 | 0.86 | 7.5 | 2.2 | 2.3 | 1.2 |
| Y2-180L-4 | 22 | 1470 | 42.6 | 91.0 | 0.86 | 7.5 | 2.2 | 2.3 | 1.2 |
| Y2-200L-4 | 30 | 1470 | 57.2 | 92.0 | 0.86 | 7.2 | 2.2 | 2.3 | 1.2 |
| Y2-225S-4 | 37 | 1480 | 69.6 | 92.5 | 0.87 | 7.2 | 2.2 | 2.3 | 1.1 |
| Y2-225M-4 | 45 | 1380 | 84.0 | 92.8 | 0.87 | 7.2 | 2.2 | 2.3 | 1.1 |
| Y2-250M-4 | 55 | 1480 | 102.9 | 93.0 | 0.87 | 7.2 | 2.2 | 2.3 | 1.0 |
| Y2-280S-4 | 75 | 1480 | 138.0 | 93.8 | 0.87 | 7.2 | 2.2 | 2.3 | 1.0 |
| Y2-280M-4 | 90 | 1480 | 165.6 | 94.2 | 0.87 | 7.2 | 2.2 | 2.3 | 1.0 |
| Y2-315S-4 | 110 | 1480 | 200.2 | 94.5 | 0.88 | 6.9 | 2.1 | 2.2 | 1.0 |
| Y2-315M-4 | 132 | 1480 | 239.1 | 94.8 | 0.88 | 6.9 | 2.1 | 2.2 | 1.0 |
| Y2-315L1-4 | 160 | 1480 | 288.0 | 94.9 | 0.89 | 6.9 | 2.1 | 2.2 | 1.0 |
| Y2-315L2-4 | 200 | 1480 | 358.9 | 95.0 | 0.89 | 6.9 | 2.1 | 2.2 | 1.0 |
| Y2-355M-4 | 250 | 1480 | 437.5 | 95.3 | 0.90 | 6.9 | 2.1 | 2.2 | 0.9 |
| Y2-355L-4 | 315 | 1480 | 547.4 | 95.6 | 0.90 | 6.9 | 2.1 | 2.2 | 0.8 |

**Table 2.14** Technical data of Y2 series 6-pole three-phase asynchronous motors.

| Type | Rated power (kW) | Rated load | | | | Stall current/ rated current | Stall torque/ rated torque | Maximum torque/ rated torque | Minimum torque/ rated torque |
|---|---|---|---|---|---|---|---|---|---|
| | | Speed (r/min) | Current (A) | Efficiency (%) | Power factor | | | | |
| Y2-90S-6 | 0.75 | 910 | 2.23 | 69.0 | 0.72 | 5.5 | 2.0 | 2.1 | 1.5 |
| Y2-90L-6 | 1.1 | 910 | 3.10 | 72.0 | 0.73 | 5.5 | 2.0 | 2.1 | 1.3 |
| Y2-100L-6 | 1.5 | 940 | 3.89 | 76.0 | 0.75 | 5.5 | 2.0 | 2.1 | 1.3 |
| Y2-112M-6 | 2.2 | 940 | 5.46 | 79.0 | 0.76 | 6.5 | 2.0 | 2.1 | 1.3 |
| Y2-132S-6 | 3 | 960 | 7.10 | 81.0 | 0.76 | 6.5 | 2.1 | 2.1 | 1.3 |
| Y2-132M1-6 | 4 | 960 | 9.30 | 82.0 | 0.76 | 6.5 | 2.1 | 2.1 | 1.3 |
| Y2-132M2-6 | 5.5 | 960 | 12.30 | 84.0 | 0.77 | 6.5 | 2.1 | 2.1 | 1.3 |
| Y2-160M-6 | 7.5 | 970 | 16.70 | 86.0 | 0.77 | 6.5 | 2.0 | 2.1 | 1.3 |
| Y2-160L-6 | 11 | 970 | 23.60 | 87.5 | 0.78 | 6.5 | 2.0 | 2.1 | 1.3 |
| Y2-180L-6 | 15 | 970 | 30.70 | 89.0 | 0.81 | 7.0 | 2.0 | 2.1 | 1.2 |
| Y2-200L1-6 | 18.5 | 970 | 37.70 | 90.0 | 0.81 | 7.0 | 2.1 | 2.1 | 1.2 |
| Y2-200L2-6 | 22 | 970 | 44.10 | 90.0 | 0.83 | 7.0 | 2.1 | 2.1 | 1.2 |
| Y2-225M-6 | 30 | 980 | 58.40 | 91.5 | 0.84 | 7.0 | 2.0 | 2.1 | 1.2 |
| Y2-250M-6 | 37 | 980 | 70.40 | 92.0 | 0.86 | 7.0 | 2.1 | 2.1 | 1.2 |
| Y2-280S-6 | 45 | 980 | 85.40 | 92.5 | 0.86 | 7.0 | 2.1 | 2.2 | 1.1 |
| Y2-280M-6 | 55 | 980 | 103.3 | 92.8 | 0.86 | 7.0 | 2.1 | 2.2 | 1.1 |
| Y2-315S-6 | 75 | 990 | 140.2 | 93.5 | 0.86 | 7.0 | 2.0 | 2.0 | 1.0 |
| Y2-315M-6 | 90 | 990 | 167.0 | 93.8 | 0.86 | 7.0 | 2.0 | 2.0 | 1.0 |
| Y2-315L-6 | 110 | 990 | 202.3 | 94.0 | 0.86 | 6.7 | 2.0 | 2.0 | 1.0 |
| Y2-315L2-6 | 132 | 990 | 242.3 | 94.2 | 0.87 | 6.7 | 2.0 | 2.0 | 1.0 |
| Y2-355M1-6 | 160 | 985 | 287.9 | 94.5 | 0.88 | 6.7 | 1.9 | 2.0 | 1.0 |
| Y2-355M2-6 | 200 | 985 | 358.4 | 94.7 | 0.88 | 6.7 | 1.9 | 2.0 | 0.9 |
| Y2-355L1-6 | 250 | 985 | 444.8 | 94.9 | 0.88 | 6.7 | 1.9 | 2.0 | 0.9 |
| Y2-355L2-6 | 315 | 990 | 585.5 | 95.0 | 0.88 | 6.7 | 1.9 | 2.0 | 0.9 |

**Table 2.15** Technical data of Y2 series 8-pole three-phase asynchronous motors.

| Type | Rated power (kW) | Rated load | | | | Stall current/ rated current | Stall torque/ rated torque | Maximum torque/ rated torque | Minimum torque/ rated torque |
|---|---|---|---|---|---|---|---|---|---|
| | | Speed (r/min) | Current (A) | Efficiency (%) | Power factor | | | | |
| Y2-100L1-8 | 0.75 | 690 | 2.34 | 71.0 | 0.67 | 4.0 | 1.8 | 2.0 | 1.3 |
| Y2-100L2-8 | 1.1 | 690 | 3.22 | 73.0 | 0.69 | 5.0 | 1.8 | 2.0 | 1.2 |
| Y2-112M-8 | 1.5 | 690 | 4.41 | 75.0 | 0.69 | 5.0 | 1.8 | 2.0 | 1.2 |
| Y2-132S-8 | 2.2 | 710 | 6.0 | 78.0 | 0.71 | 6.0 | 1.8 | 2.0 | 1.2 |
| Y2-132M-8 | 3 | 710 | 7.6 | 79.0 | 0.73 | 6.0 | 1.8 | 2.0 | 1.2 |
| Y2-160M1-8 | 4 | 720 | 10.0 | 81.0 | 0.73 | 6.0 | 1.9 | 2.0 | 1.2 |
| Y2-160M2-8 | 5.5 | 720 | 13.3 | 83.0 | 0.74 | 6.0 | 2.0 | 2.0 | 1.2 |
| Y2-160L-8 | 7.5 | 720 | 17.8 | 85.5 | 0.75 | 6.0 | 2.0 | 2.0 | 1.2 |
| Y2-180L-8 | 11 | 730 | 24.9 | 87.5 | 0.76 | 6.6 | 2.0 | 2.0 | 1.1 |
| Y2-200L-8 | 15 | 730 | 33.3 | 88.0 | 0.76 | 6.6 | 2.0 | 2.0 | 1.1 |
| Y2-225S-8 | 18.5 | 730 | 40.1 | 90.0 | 0.76 | 6.6 | 1.9 | 2.0 | 1.1 |
| Y2-225M-8 | 22 | 740 | 46.8 | 90.5 | 0.78 | 6.6 | 1.9 | 2.0 | 1.1 |
| Y2-250M-8 | 30 | 740 | 63.0 | 91.0 | 0.79 | 6.6 | 1.9 | 2.0 | 1.1 |
| Y2-280S-8 | 37 | 740 | 76.2 | 91.5 | 0.79 | 6.6 | 1.9 | 2.0 | 1.1 |
| Y2-280M-8 | 45 | 740 | 92.5 | 92.0 | 0.79 | 6.6 | 1.9 | 2.0 | 1.0 |
| Y2-315S-8 | 55 | 740 | 110.4 | 92.8 | 0.81 | 6.6 | 1.8 | 2.0 | 1.0 |
| Y2-315M-8 | 75 | 740 | 148.1 | 93.0 | 0.81 | 6.6 | 1.8 | 2.0 | 0.9 |
| Y2-315L1-8 | 90 | 740 | 177.6 | 93.8 | 0.82 | 6.6 | 1.8 | 2.0 | 0.9 |
| Y2-315L2-8 | 110 | 740 | 215.8 | 94.0 | 0.82 | 6.4 | 1.8 | 2.0 | 0.9 |
| Y2-355M1-8 | 132 | 730 | 256.8 | 93.7 | 0.82 | 6.4 | 1.8 | 2.0 | 0.9 |
| Y2-355M2-8 | 160 | 730 | 307.8 | 94.2 | 0.82 | 6.4 | 1.8 | 2.0 | 0.9 |
| Y2-355L2-8 | 200 | 730 | 383.0 | 94.5 | 0.83 | 6.4 | 1.8 | 2.0 | 0.9 |
| Y2-355L4-8 | 250 | 740 | 490.0 | 94.9 | 0.82 | 6.8 | 1.8 | 2.0 | 0.9 |
| Y2-400L1-8 | 315 | 740 | 613.0 | 95.0 | 0.82 | 6.8 | 1.8 | 2.0 | 0.9 |

**Table 2.16** Categories, characteristics, and applications of main series products deriving from three-phase asynchronous motors.

| Serial number | Series name | Type | Performance or structural characteristics | Applications |
|---|---|---|---|---|
| 1 | Asynchronous motors | Y, Y2 | Most are squirrel cage three-phase asynchronous motors, protection grade includes IP23, IP44, IP54, cast-iron frame, and cast-aluminum rotor. | Used for driving pumps, fans, compressors, machine tools, and other general mechanical equipment. |
| 2 | Efficient asynchronous motors | YX | Efficiency is 3% higher than ordinary asynchronous motors, good magnetic materials are used to increase the usage of effective material. | Used in high load factor, long running time occasion. |
| 3 | Winding asynchronous motors | YR | Enclosed type, protective housing, and wound rotor. Reducing the starting current and increasing the starting torque can be achieved by connecting external resistor. | Mainly used for metallurgy auxiliary equipment and general hoisting machinery. |
| 4 | High slip asynchronous motors | YH | The rotor uses small groove and deep groove, casted with high resistance aluminum alloy, having higher starting torque, smaller starting current, bigger slip, and soft mechanical characteristics. | Used for the loads with greater flywheel torque and strong impact, mechanical equipment with several times start and reverse. Such as shear, punch press, and so on. |
| 5 | Multispeed asynchronous motors | YD | Changing the poles by changing stator winding connection methods can get more kinds of speed. The regulation method is sample and the efficiency is high. | Mainly used for mechanical equipment that requires 2–4 kinds of hierarchical speed, such as fan pumps, machine tools, dyeing machines, and so on. |
| 6 | High starting torque asynchronous motors | YQ | The rotor adopts double cage or deep groove. The starting torque is huge. | Applied to start heavy static load and large inertia machine, such as compressor, crusher, and so on. |
| 7 | Explosion-proof safe asynchronous motors | YA | Does not produce sparks, arc and dangerous temperature during normal operation. | Only used in the occasion that only form an explosive mixture in abnormal situations. |
| 8 | Flameproof asynchronous motors | YB | Motors structure takes explosion-proof measures into account; it has the enclosed self-fan cooled type; and it enhances the shell mechanical strength. | Used in the place filled with explosive mixture formed by combustible gas or steam in mixture with air. |
| 9 | Asynchronous motors for crane and metallurgical uses | YZ, YZR | Intermittent duty; enclosed self-fan cooled type; use cage rotor casted with high resistance aluminum alloy, high starting torque, which can start frequently, large overload capacity, high slip. YZR is wound rotor type structure, emphasizing on high reliability. | Applied to general hoisting equipment and metallurgical equipment. Have large starting torque and small range of speed regulation characteristics. |

(*Continued*)

**Table 2.16** (Continued)

| Serial number | Series name | Type | Performance or structural characteristics | Applications |
|---|---|---|---|---|
| 10 | Torque asynchronous motors | YLJ | Soft mechanical characteristics, running stably in the range of synchronous speed. | Drive the constant tension, constant linear velocity (winding) and constant torque mechanical equipment. |
| 11 | Variable frequency speed control asynchronous motors | YVF2 | Meet the requirements for the power supply of inverter, and fan is adopted independent small power motors to meet the requirements of heat radiation. | Applied to nonpole speed-adjusting mechanical equipment. |
| 12 | Low-noise and low vibration asynchronous motors | YZC | Use special low noise motors bearings, improve the machining accuracy and rotor balance precision, modify the electromagnetic parameters, and improve the overall machines level. | Mainly used for precision machine tools. |
| 13 | Gear head asynchronous motors | YCJ | Composed of Y series motors coupled directly with gear reducer. Reducer uses outer gearing, involute cylindrical gear, which can transmit power from positive and negative two directions. | Dedicated to low-speed high torque machines, suitable for mining, steel rolling, and chemical and other industrial equipment. |
| 14 | Electromagnetic braking asynchronous motors | YEP | The motors working mode is S3, and the duty cycle (FC) is 25%. the rotor non-axial end is equipped with magnetic blocks and brake device, composited with motors, having brake fast, compact structure characteristics. | Used in starting transport machinery, lifting machinery and other occasions for prompt and accurate parking. |
| 15 | Asynchronous motors for elevator | YTD | Short-time duty variable frequency control motors, or consequent-poles motors. | Drive lift. |
| 16 | Asynchronous motors for marine | Y-H | High strength shell, which can withstand the impact vibration and turbulence. Insulation treatment can deal with the corrosion of condensation, salt fog and mould. | Used for the general ships of sea or river. |
| 17 | Submersible asynchronous motors for deep water well | YQS2 | Small outer diameter size, slender motors, the interior adopts seal structure inflated by water. The wire uses water-fast enameled wire, without foots; frame is fixed with flange. | Used for submersible pump supports, applied to lift water in industry and agriculture. |
| 18 | Submersible asynchronous motors for deep oil well | YQY | Small outer diameter size, slender motors, the interior adopts seal structure inflated by oil. Using high-temperature-resistant wire composited with submersible pump. | Used in oil field production equipment. |

Understanding the categories, characteristics, and functions of commonly used three-phase asynchronous motors can help make good decision in choosing motor series. The motor must be suitable for its application requirements to fulfill the purpose of energy saving. In the case of choosing a series motor for the tower cranc equipment, if we choose Y series, due to the high starting torque and frequent start, a large-capacity Y series is needed, which would appear "big horse pull a small carriage" phenomenon, resulting in a waste of energy.

## 2.9 Discussion Topics in the Chapter

1) What are the advantages and disadvantages of three-phase asynchronous motors?
2) What is the power range of small- and medium-type three-phase asynchronous motors?
3) What is the meaning of three-phase asynchronous motors model number (such as Y-225M-6)?
4) Among the commonly used three-phase asynchronous motors, what are characteristics and applications of Y Y2 YX YR YH series motors?
5) What is the slip range of three-phase asynchronous motors in electric operation? What is the slip range of three-phase asynchronous motors in the generation operation?
6) How many kinds of losses do the three-phase asynchronous motors have? Try to draw out the power balance figure of the three-phase asynchronous motors.
7) What curves do the operating characteristic of the three-phase asynchronous motors refer to? What can we get from the curves?
8) Draw out the mechanical characteristic curves of the three-phase asynchronous motors, and explain their characteristics.
9) What are the meanings of the minimum allowable values and efficiency class?
10) Know about the technology data of Y2 series three-phase asynchronous motors, and learn to refer to the product technical information.

# 3

# Economic Operation of the Three-Phase Induction Motor

The key to study induction motor energy saving is its economic operation. First, this chapter not only analyzes all kinds of loss of the induction motor and the method of reducing loss but also studies the operating efficiency of the induction motor and its power factor. Second, the chapter introduces the standard of "Economic Operation of the Induction Motor" that includes the terms and definitions of the induction motor economic operation, the basic requirements of economic operation, the calculation method of the comprehensive efficiency, and the judgment for economic operation of induction motor. Last, according to the different actual situation, the calculation and evaluation methods of energy-saving effect for the induction motor are introduced.

## 3.1 Loss Analysis of the Three-Phase Induction Motor

The loss of the induction motor includes iron loss, stator copper loss, rotor copper loss, mechanical loss, and stray loss. Iron loss and mechanical loss are often referred to as the constant loss and stator copper loss, rotor copper loss, and stray loss are called the variable loss. In terms of the motor itself, energy saving is to reduce the loss and improve its efficiency. Reducing loss should mainly focus on main loss. Taking Y series motor as an example, for a small power motor copper loss and iron loss is the main proportion, so the motor should appropriately increase its effective materials. For instance, increase the conductor cross-section to reduce the stator copper loss and improve efficiency. For high power motor, the main loss is mechanical loss and stray loss, so that it should reduce air friction loss and stray loss through various measures. The detailed analysis of various losses is as follows.

### 3.1.1 The Analysis of Iron Loss

#### 3.1.1.1 Iron Loss

Iron loss ($P_{Fe}$) is the eddy current loss and hysteresis loss that is caused by the alternating changing magnetic field in the motor iron core. In normal operation, the slip and the frequency of the changing magnetic field in the rotor core is very small, generally the frequency is only 1–3 Hz/s, so we ignore the rotor iron loss in iron loss. Iron loss should also contain the no-load stray losses. No-load stray loss is the loss caused by the leakage

*Energy-saving Principles and Technologies for Induction Motors*, First Edition. Wenzhong Ma and Lianping Bai.

of magnetic flux when no-load current passes through the stator windings in the stator frame, end cover, and other metals. Because the no-load current is approximate constant, the loss is constant. No-load stray loss accounts for only about 1% of the iron loss, it can also be ignored. The iron loss depends on the material performance of the motor iron core, frequency, and magnetic flux density. Iron loss is proportional to the square of the magnetic flux density. According to the theory of the motor,

$$P_{Fe} = KG_{Fe} f^{1.3} B^2 \tag{3.1}$$

where $K$, the correction coefficient of iron loss;
$G_{Fe}$, the weight of the stator core;
$f$, power frequency;
$B$, magnetic flux density.

From Equation (3.1), it can be seen that iron loss is proportional not only to the square of the magnetic flux density but also to the weight of the stator core. For the motors with same power, the larger the pole number and the greater the volume of the motors, the bigger the iron loss they bring. For example, from Tables 2.13, 2.14, and 2.15, it can be found that the 30 kW/8-pole motor's frame size is 250, 30 kW/6-pole motor's frame size is 225, and 30 kW/4-pole motor's frame size is 200. It is obvious that the larger the pole number and the larger the size of motors, the greater the iron loss.

For the 15–75 kW induction motor of common use, the iron loss accounted for 20–30% of the total loss.

#### 3.1.1.2 The Methods to Reduce Iron Loss

Iron loss is composed of two parts: hysteresis loss and eddy current loss. The use of cold-rolled silicon steel sheet that has good permeability can reduce hysteresis loss. For a given thickness of steel sheet and a certain magnetic flux density, hysteresis loss depends on the quality of the material. High-quality silicon steel is expensive, so we have to make a comparison between the iron loss reducing and the cost increasing. Now it has stipulated that we must use the cold-rolled silicon steel sheets instead of the hot-rolled silicon steel sheets in China.

Using thin silicon can reduce the eddy current loss. Reducing the magnetic flux density can reduce the hysteresis loss and eddy current loss. To reduce the winding current we can reduce the air gap length. For the manufacture and installation reasons, it is ineffective. Enlarging the core can effectively reduce the magnetic flux density, so we often increase the iron core's length to reduce the magnetic flux density. Generally, the high efficiency motor uses 35% iron more than the same power motors.

### 3.1.2 The Analysis of Mechanical Loss

#### 3.1.2.1 Mechanical Loss

The mechanical loss includes bearing friction loss and ventilation system loss, wound-rotor also includes brush friction loss. Bearing friction loss is mainly related to the bearing model, the level of installation, and the grease. The ventilation system loss mainly depends on the material of the fan, the efficiency of fan, and the rationality of the air duct's design. The mechanical loss is also related to the motor speed. Bearing friction loss is proportional to the square of the motor speed and the ventilation loss is proportional to the cube of the motor speed.

For the 15–75 kW/4-pole induction motors, mechanical loss occupies 14–16% of total loss.
For the 15–75 kW/6-pole induction motors, mechanical loss occupies 12–14% of total loss.
For the 15–75 kW/8-pole induction motors, mechanical loss occupies 10–12% of total loss.

Due to the reduction of the wind friction loss, for the motors with same power, the larger the number of the pole (speed-drop), the lesser the air friction loss. For the motor that has the same number of poles, the larger the capacity, the greater the proportion of air friction loss in the total loss. That is mainly because of cooling needs.

#### 3.1.2.2 The Methods to Reduce Mechanical Loss

The mechanical loss is caused by the motor friction and ventilated system. Friction mainly refers to the bearing friction; using the high-quality bearing and high-quality lubricating oil can reduce the mechanical loss. The wind friction is caused by cooling air and the rotating part of motor, so using high efficiency fan and the reasonable design of the ventilation structure can reduce the friction loss.

### 3.1.3 Stator and Rotor Copper Loss Analysis

#### 3.1.3.1 Stator and Rotor Copper Loss

Stator and rotor copper loss ($P_{Cu1}$, $P_{Cu2}$) is also known as the load loss. Copper loss is referred to the loss that is generated when the stator and rotor current run through the winding resistance. Copper loss depends on the load current and the winding resistance. It can be known from the equivalent circuit that $P_{Cu1} = 3I_1^2 r_1$, and $P_{Cu2} = 3I_2^2 r_2$. For the 15–75 kW induction motor, copper loss occupies 40–60% of the total loss. With the increasing of motor capacity, the proportion of copper loss decreases.

#### 3.1.3.2 The Measures to Reduce Stator and Rotor Copper Loss

When the current passes through the winding, it generates heat, and this is the winding copper loss; the best way to reduce the copper loss ($I^2R$) is to reduce the resistance. Under the allowance of the slot full rate, we can increase the stator winding wire's section area to reduce the winding resistance. But it is restrained by the size of the motor and the cost.

Squirrel cage induction motor can appropriately increase the conducting bar's sectional area to reduce the rotor resistance. For wound induction motor, we can use copper wires and appropriately increase the cross-sectional area of the rotor copper wires.

Furthermore, reducing the air gap between the stator and the rotor, adopting a high-quality silicon steel sheet, reducing the magnetic flux density appropriately, and reducing exciting current can reduce the stator current and copper loss and also improve the power factor of the motor.

### 3.1.4 The Analysis of Stray Loss

#### 3.1.4.1 Stray Loss

Stray loss ($P_s$) is also known as additional loss. Stray loss is mainly composed of the copper loss caused by leakage of magnetic flux in the winding, iron core, shaft, end cover, and other metal parts; pulsation loss caused by cogging effect; and copper loss caused by high harmonics in the stator and rotor.

For the 15–75 kW/8-pole induction motors, stray loss occupies about 15–25% in the total loss.

#### 3.1.4.2 The Measures to Reduce Stray Loss

Stray loss includes the copper loss that is caused by leakage of magnetic flux, the pulse vibration loss that is caused by cogging effect, surface loss, and the loss that is caused by higher harmonics. The main methods to reduce the stray loss are as follows: using series of sine windings and distributed windings to reduce higher harmonics, selecting the appropriate number of slots for the stator and rotor, designing the rotor slot shape for chute, choosing the best gap length, using magnetic slot wedges, and so on.

### 3.1.5 The Power Grid Quality's Impact on the Loss

Power quality has a direct impact on the motor loss. For example, the voltage fluctuation, the imbalance of three-phase voltage, and the high harmonics in the grid can increase the loss of induction motor.

#### 3.1.5.1 The Influence of Voltage Fluctuation on Various Losses

1) The influence on iron loss

$$P'_{\mathrm{Fe}} \approx P_{\mathrm{Fe}}\left(\frac{U'}{U_N}\right)^2 \tag{3.2}$$

As can be seen from the Equation (3.2), when power supply voltage rises, the iron loss increases. When the power supply voltage drops, the iron loss reduce. However, the output torque is proportional to the square to decrease, rotating speed drops, the current increases, the copper loss increases, and the total loss increases.

2) The influence on mechanical loss

$$P'_{\mathrm{fw}} \approx P_{\mathrm{fw}}\left(\frac{1-s}{1-s_{\mathrm{N}}}\right)^2 \tag{3.3}$$

When the power supply voltage fluctuates (for instance, ±5%), rotating speed changes little and has little impact on mechanical loss.

3) The influence on stator copper loss

$$P'_{\mathrm{Cu1}} \approx P_{\mathrm{Cu1}}\left\{\left(\frac{I'_0}{I'_{\mathrm{N}}}\right)^2 + \left[1-\left(\frac{I'_0}{I'_{\mathrm{N}}}\right)^2\right]\left[\frac{(1-s_{\mathrm{N}})s}{(1-s)s_{\mathrm{N}}}\right]^2\right\} \tag{3.4}$$

With a constant external load torque, when the supply voltage increases, the excitation current of stator increases, the speed increases, and the stator copper loss increases slightly.

4) The influence on rotor copper loss

$$P'_{\mathrm{Cu2}} \approx P_{\mathrm{Cu2}}\left(\frac{U_{\mathrm{N}}}{U'}\right)^2\left(\frac{1-s_{\mathrm{N}}}{1-s}\right)^2 \tag{3.5}$$

With a constant external load torque, when the supply voltage increases, the speed increases, the slip ratio decreases slightly, and the rotor copper loss decreases slightly.

5) The influence on stray loss

$$P'_s \approx P_s\left(\frac{U_{\mathrm{N}}}{U'}\right)^2\left(\frac{1-s_{\mathrm{N}}}{1-s}\right)^2 \tag{3.6}$$

When the supply voltage increases, like the rotor copper loss, the stray loss reduces. $P'_{\mathrm{Fe}}, P_{\mathrm{Fe}}, P'_{\mathrm{fw}}, P_{\mathrm{fw}}, P'_{\mathrm{Cu1}}, P_{\mathrm{Cu1}}, P'_{\mathrm{Cu2}}, P_{\mathrm{Cu2}}, P'_s, P_s$ correspond to all kinds loss of $U'$ or $U_{\mathrm{N}}$. From Equation (3.6), it can be seen that when the power supply voltage increases, the total loss has an increasing trend; when the grid voltage drops, the iron loss of motor reduces by the square of the voltage. When the load is constant, the operating voltage drops, the motor torque decreases and slip becomes larger, and the stator and rotor copper loss will increase. Therefore, in the case of a motor running in a constant load, if the voltage drops, the total loss increases. Only in the case of very low load rate, when the voltage drops, the total loss of the motor will drop and the efficiency will increase.

#### 3.1.5.2 The Unbalance of the Three-Phase Voltage's Effect on Loss

In three-phase power system, the three-phase voltage and current are unbalanced in a different degree. Unbalanced three-phase voltage directly affects the loss and performance of a motor. The definition of national standards (GB/T15543-2008) for three-phase voltage unbalance is that three-phase voltage in different amplitude or phase difference is not 120 degrees, or both. For unbalance of three-phase voltage, decompose it into positive, negative, and zero sequence components and then analyze the unbalance according to the relative relation of negative and zero sequence component and the positive sequence component. The unbalance of three-phase voltage is defined as follows:

$$\varepsilon_{\mathrm{U2}} = \frac{U_2}{U_1} \times 100\% \tag{3.7}$$

where $\varepsilon_{\mathrm{U2}}$, the unbalance of negative sequence voltage;

$U_1$, the root mean square value of the positive sequence component of three-phase voltage (V);

$U_2$, the root mean square value of the negative sequence component of three-phase voltage (V).

In the national standard, during normal operation of the grid, the unbalance of negative sequence voltage is not more than 2%. If the unbalance of negative sequence voltage exceeds the national standard limited value, it will have a large influence on the motor loss. The analysis shows that, in the motor winding, a small component of negative sequence voltage will cause a larger negative sequence current. For instance, when the component of negative sequence voltage reaches 5%, the negative sequence current will reach 20–35% of the rated current, which will cause great loss and overheat in the winding.

Negative sequence current will produce negative sequence magnetic field and the torque generated by a negative sequence magnetic field is negative, which will generate a braking torque. Because the mechanical power produced by the negative sequence component is negative, it will absorb some of the mechanical power, so the rotor load current increases and the copper loss of rotor will increase.

The negative sequence component not only causes additional iron loss, copper loss, and stray loss in the rotor but also increases the losses in the stator. If there are 3.5% unbalanced voltage in the motor, then the total loss of motor will be increased by 20%, efficiency will be decreased by 3–4%.

In short, the induction motors will deteriorate its performance when voltage is unbalanced, overload ability and efficiency drops, and there is a risk of local overheating.

For zero sequence component unbalance, the national standard has been defined, but there is no limit. The reason is that zero sequence components can be suppressed by the connection way of windings.

Thus, in the monitoring of energy saving of induction motors, we should pay attention to the three-phase voltage unbalance.

#### 3.1.5.3 The Impact of Higher Harmonic Current on the Induction Motor Loss

The grid has many users of electrical equipment which is a nonlinear load. The load will generate harmonic current and harmonic voltage, which will inject into the grid, resulting in higher harmonic voltage of corresponding frequency in the system impedance; thus the system voltage wave form will be distorted. If the higher harmonic source capacity is large, the harmonic effects cannot be ignored.

If there is a harmonic current in the three-phase asynchronous motor, it will produce a higher frequency rotating magnetic field, increasing the stray loss, overheating the alveoli, and even burning the motor.

## 3.2 Efficiency and Power Factor of the Three-Phase Asynchronous Motor

The efficiency and the power factor are important technical indicators of induction motor. The efficiency reflects the size of loss of motor itself and the power factor reflects the loss which is caused by the motor's reactive current on the line. The key to the economic operation of an induction motor is to control the motor operational efficiency and the power factor; thus in the process of operation the motor can achieve a higher operational efficiency and power factor. See the following for an in-depth research on motor efficiency and power factor.

### 3.2.1 The Definition of Induction Motor's Efficiency and Power Factor

#### 3.2.1.1 The Definition of the Induction Motor's Efficiency

The efficiency of the induction motor is referred to as the ratio of output power and input power, that is

$$\eta = \frac{P_2}{P_1} \times 100\% \tag{3.8}$$

where $P_2$, the output power of motor (kW);
$P_1$, the input power of motor (kW).

In practical applications, there are rated efficiency, operational efficiency, and the highest efficiency for a motor. Rated efficiency is the efficiency when the output of the motor is a rated value, generally represented using $\eta_N$. The running efficiency of the motor is the efficiency when the motor is running in one load, which is a point on the efficiency curve, for short as the operational efficiency, often represented using $\eta$. If the motor runs at the highest point of the efficiency curve, this point is called the motor maximum efficiency point, and we often use $\eta_m$ to represent it.

#### 3.2.1.2 The Definition of the Induction Motor's Power Factor

The power factor of the three-phase induction motor is the ratio of the motor's input active power and its apparent power, that is,

$$\cos\varphi = \frac{P_1}{\sqrt{3}U_1 I_1} \tag{3.9}$$

where $\cos\varphi$, power factor of the induction motor;
$P_1$, the input active power of the motor (kW);
$U_1$, the line voltage of the motor input end (kV);
$I_1$, the line current of the motor input end (A).

The power factor of the motor at rated load is called the rated power factor $\cos\varphi_N$, at no-load is called no-load power factor $\cos\varphi_0$, and at one load is represented as $\cos\varphi$.

### 3.2.2 The Calculation of Efficiency and Power Factor of Induction Motors

From the point of energy saving, we pay more attention on the operational efficiency $\eta$ and power factor $\cos\varphi$ of the induction motor. Through the analysis of the motor efficiency $\eta = f(\beta)$ and power factor $\cos\varphi = f(\beta)$ (where $\beta$ is the ratio of output power and rated power), to find out the method to improve the operational efficiency and power factor. Then we can improve the economic operation level of the motor and accomplish the goal of energy saving.

#### 3.2.2.1 The Calculation of Operation Efficiency of the Induction Motor

The running efficiency of three-phase induction motors can be obtained from the formula $\eta = (P_2/P_1)\times 100\%$. In operation, the input power subtracting the output power is the active loss, that is $(P_1 - P_2)$, it is also the total loss of motor in operation. That is,

$$\Delta P = P_{Fe} + P_{Cu1} + P_{Cu2} + P_{fw} + P_s \tag{3.10}$$

Testing various losses in the process of motor running is very difficult; we can first measure the load coefficient according to the economic operation standard of induction motor (GB/T12497-2006); we can then calculate the total active loss of the running motor.

$$\Delta P = \Delta P_0 + \beta^2\left(\Delta P_N - \Delta P_0\right) \tag{3.11}$$

where $\Delta P$, the total active loss of the motor in operation (kW);
$\Delta P_0$, the active loss of the motor at no-load (kW);
$\Delta P_N$, the active loss of the motor at rated load (kW);
$\beta$, load coefficient, that is

$$\beta = \frac{P_2}{P_N} \tag{3.12}$$

where $P_2$, the output power of the motor (kW);
$P_N$, the rated power of the motor (kW).

The load coefficient multiplied by 100% is known as the load rate, which is usually represented by $\beta$.

Active power loss of the motor at rated load can be determined by rated efficiency and rated power, that is

$$\Delta P_N = \left(\frac{1}{\eta_N} - 1\right) P_N \tag{3.13}$$

where $\eta_N$ is the rated motor efficiency and $P_N$ and $\eta_N$ are the data given on the plate name. So the total active power loss in operation is

$$\Delta P = \Delta P_0 + \beta^2 \left[\left(\frac{1}{\eta_N} - 1\right) P_N - \Delta P_0\right] \tag{3.14}$$

Hence, the operational efficiency of the motor can be calculated by the following formula:

$$\eta = \frac{P_2}{P_1} = \frac{\beta P_N}{\beta P_N + \Delta P} = \frac{\beta P_N}{\beta P_N + \Delta P_0 + \beta^2 \left[\left(\frac{1}{\eta_N} - 1\right) P_N - \Delta P_0\right]} \tag{3.15}$$

From the above formula, we can learn that operational efficiency of the motor is associated with $\eta_N$, $\Delta P_0$, and $\beta$. $\eta_N$ and $\Delta P_0$ are determined by the characteristics of the motor. For a certain motor, $\eta_N$ and $\Delta P_0$ are fixed values. Therefore, the operational efficiency of a certain motor depends on the motor load rate $\beta$.

#### 3.2.2.2 The Calculation of Operational Power Factor of the Induction Motor

The power factor of a running induction motor is

$$\cos\varphi = \frac{P_1}{\sqrt{3}U_1 I_1} = \frac{P_1}{\sqrt{P_1^2 + Q_1^2}} \tag{3.16}$$

In the above formula,

$$P_1 = \beta P_N + \Delta P_0 + \beta^2 \left[\left(\frac{1}{\eta_N} - 1\right) P_N - \Delta P_0\right] \tag{3.17}$$

$$Q_1 = Q_0 + \beta^2 \left[ Q_N - Q_0 \right] \tag{3.18}$$

$$Q_0 = \sqrt{3U^2 I_0^2 \times 10^{-6} - \Delta P_0^2} \tag{3.19}$$

If we substitute $Q_N = \left( P_N / \eta_N \right) \tan \varphi_N$ into (3.18), we can learn that

$$Q_1 = Q_0 + \beta^2 \left[ \frac{P_N}{\eta_N} \tan \varphi_N - Q_0 \right] \tag{3.20}$$

If we substitute formulas (3.17) and (3.20) into (3.16), we can learn that

$$\cos\varphi = \frac{\beta P_N + \Delta P_0 + \beta^2 \left[ \left( \frac{1}{\eta_N} - 1 \right) P_N - \Delta P_0 \right]}{\sqrt{\left\{ \beta P_N + \Delta P_0 + \beta^2 \left[ \left( \frac{1}{\eta_N} - 1 \right) P_N - \Delta P_0 \right] \right\}^2 + \left\{ Q_0 + \beta^2 \left[ \frac{P_N}{\eta_N} \tan \varphi_N - Q_0 \right] \right\}^2}} \tag{3.21}$$

From formula (3.19), we can learn that the power factor of the motor $\cos\varphi$ has relationship with $I_0, \Delta P_0, P_N, \eta_N, \cos\varphi_N, \beta$. For one specified motor, the size of $\cos\varphi$ only has a relationship with load coefficient $\beta$. In the case where the motor's rated parameters are known, the load factor can be calculated by the following formula:

$$\beta = \frac{-P_N/2 + \sqrt{P_N^2 / 4 + \left( \Delta P_N - \Delta P_0 \right)\left( P_1 - \Delta P_0 \right)}}{\Delta P_N - \Delta P_0} \times 100\% \tag{3.22}$$

where $P_N$, the rated power (kW);
$\Delta P_N$, the rated total loss (kW);
$\Delta P_0$, the loss at no-load (kW);
$P_1$, the input power (kW).

In practical work, if we want to know the size of load rate, we should first obtain the output power $P_2$, which is not easy to measure directly. In the above formula, $P_N$, $\Delta P_N$, and $\Delta P_0$ are known parameters. On the spot, if we can know the input power $P_1$, we can know the load rate of the motor.

### 3.2.3 The Efficiency and Power Factor Curve of the Induct Motor

#### 3.2.3.1 The Power Factor Curve of the Motor and Its Drawing

If we know the parameters $I_0, \Delta P_0, P_N, \eta_N, \cos\varphi_N$ of a motor, we can draw the efficiency and power factor curve based on formulas (3.15) and (3.19), that is $\eta = f(\beta)$ and $\cos\varphi = f(\beta)$. Furthermore, according to the motor efficiency and power factor curve we can analyze the running state of the motor and determine whether the motor is on an economic operating state.

**Example 3.1** There is an induction motor, type Y2-250M-4. From Table 2.13 we can learn that its power is 55 kW, 4 poles, rated efficiency 0.93, rated power factor 0.87, no-load reactive power 20.02 kvar, and no-load loss 1.675 kW. Get its function expression of the efficiency and the power factor, and draw the curve $\eta = f(\beta)$ and $\cos\varphi = f(\beta)$.

**Solution**

The rated active loss

$$\Delta P_N = \left(\frac{1}{\eta_N} - 1\right) P_N = \left(\frac{1}{0.93} - 1\right) \times 55 = 4.14\,\text{kW}$$

The rated reactive power

$$Q_N = \frac{P_N}{\eta_N} \tan\varphi_N = \frac{55}{0.93} \times 0.5667 = 33.516\,\text{kvar}$$

Substituting the above data into formula (3.15), we can obtain the operating efficiency expression

$$\eta = \frac{55\beta}{2.465\beta^2 + 55\beta + 1.675} \tag{3.23}$$

When the motor is running at rated load, we can learn that $\eta = 0.93 = \eta_N$ with the condition which $\beta$ equals 1 in the formula (3.23).

Substituting the above data into formula (3.19), we can obtain the power factor expression of the running motor.

$$\cos\varphi = \frac{2.465\beta^2 + 55\beta + 1.675}{\sqrt{188.22\beta^4 + 271.15\beta^3 + 3573.64\beta^2 + 184.25\beta + 403.61}} \tag{3.24}$$

When the motor is running at the rated load, we can know that $\cos\varphi = 0.87 = \cos\varphi_N$ with $\beta = 1$ in formula (3.24).

If we take $\beta$ as an independent variable, we can draw Figure 3.1 by formulas (3.23) and (3.24).

#### 3.2.3.2 The Analysis of Efficiency Curve and Power Factor Curve

1) The efficiency curve of the induction motor has a wide high efficiency area. Generally if the load rate of series Y and Y2 motor is 35–40%, the motor efficiency has a decline trend. If the load rate is less than 30%, the efficiency begins to decline significantly. So as long as the load rate of running motor is not very low, the motor efficiency is generally higher.
2) Power factor value is very small at no-load, generally below 0.1, and increases with the increasing load rate. If load rate reaches about 60%, it is close to the rated value. For motors of the same capacity, the power factor curve for a motor that has more poles rises faster. A small capacity motor's power factor curve rises faster than the large capacity motor.
3) The efficiency of different series, different types of motors, and their power factors are not the same. Generally speaking, for the same capacity of squirrel cage motor

and the winding type motor, squirrel cage motor's efficiency and power factor is higher; for a high speed motor, its efficiency and power factor is higher than a low speed motor; for the same type motor, the more capacity, the higher the efficiency and power factor.

4) For the same motor, its efficiency curve is not invariable. Too much time in service and poor maintenance will increase all kinds of loss and the efficiency will decline.

In summary, in order to make the motor operate economically, the motor must be chosen in an appropriate type and capacity, so that mechanical characteristics of the motor match the load and strive to have the highest operating efficiency. For the motor in service we should improve the load rate as possible. Strengthen the maintenance and overhaul, take various reform measures to reduce loss, and improve the efficiency of the running motor.

## 3.3 Economic Operation of the Three-Phase Induction Motor

The efficiency curve has a wide area with high efficiency, as is shown in Figure 3.1; as long as the load rate is not less than 20%, the efficiency is high. But the load rate of 20% cannot be selected; when the load rate was 20%, the power factor was only 0.53. When the motor absorbs reactive power from the grid, the active power loss of the grid will increase. The efficiency and power factor should be considered when choosing the load rate of motor. Therefore, the state issued the standard of "economic operation of the three-phase induction motors" (GB/T 12497-2006) in 2006,

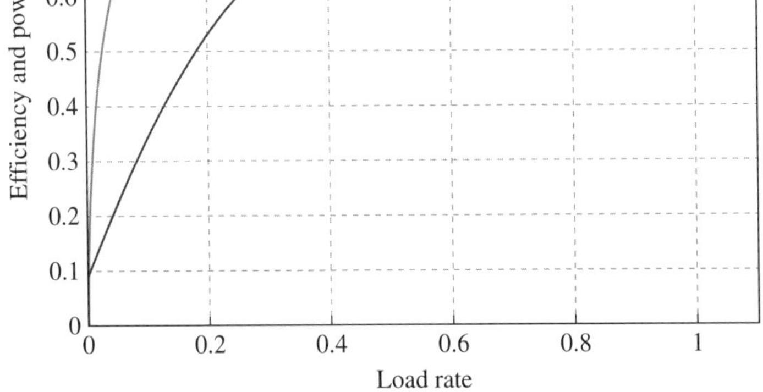


**Figure 3.1** The efficiency and power factor curve of Y2-250M-4 motor.

the standard made detailed provisions on the economic operation of the three-phase induction motors. The main contents include eight aspects: terms and definitions of economic operation of motors, the principle and technological conditions of economic operation of motors, the weighted average comprehensive efficiency and the overall economic load rate of the motor, motor replacement or renovation, running state test of the motor, calculation and judgment of economic operation, management of motor economic operation, and key motor plan. The standard of "Economical Operation of a Three-phase Induction Motor" involves a lot of content. The chapter only describes the terms and definitions of the economic operation of the motor, principles and technical conditions of the economic operation of the motor, the calculation of motor operating efficiency and economic load rate, as well as the judgment of the economic operation. In the following chapters, motor running tests, test conditions, and motor reactive power compensation on the spot will be introduced. As for the remaining contents, please refer to the original standard text or other reference information.

The standard of "economic operation of the three-phase induction motors" is an important basis for motor energy conservation. For the motor manufacturers, users, and technical supervision departments, they should all earnestly study and master the standard.

### 3.3.1 The Terms and Definitions of Economic Operation for the Three-Phase Induction Motor

1) Economic operation of the motor on the premise that the operating characteristics requirements of the drag load are met, safe and reliable, and does not affect the production, does not have a negative environmental impact, the operation mode of saving energy and maintenance costs.
2) Economic equivalent of reactive power—the active power loss of grid when the motor consumes 1 kvar reactive power.
3) Comprehensive power loss of motor—the sum of active power loss and active power loss of the grid caused by the increase of reactive power.
4) Comprehensive power consumption of motor—the sum of output power of the motor and corresponding comprehensive power loss.
5) The comprehensive efficiency of motor—the ratio of output power of the motor to corresponding comprehensive power consumption.
6) The rated comprehensive efficiency of motor—the comprehensive efficiency of the motor at the rated load operation.
7) Load factor—the ratio of the motor output power to its rated power is called load rate when the load factor is expressed as a percentage.
8) Economic load rate—the load rate of motor when its efficiency is the highest.
9) Comprehensive economic load rate—load rate of motor when its comprehensive efficiency is the highest.
10) Weighted average comprehensive efficiency—in a certain time period, the weighted average comprehensive efficiency of the motor running at a different load rate.

### 3.3.2 Basic Requirements for Economical Operation of the Three-Phase Induction Motor

1) We should be fully aware of the characteristics of the mechanical load (hereinafter referred to as load) being dragged before motor selection. If the load has no special requirements on start-up, braking, and speed control, we should use a squirrel cage induction motor. From the aspect of energy saving, motors complying with the efficiency standards in Table 2.9 should first be selected, and products eliminated by the state should not be chosen. If the load has special requirements on the start, brake, and speed regulation, motor selection should meet the locked-rotor torque and maximum torque requirements; the motors selected should be reasonable matching the speed control mode.

   We should also consider the work environment of the motor, load characteristic, working mode, installation, and so forth. Choosing the motors reasonably is the necessary condition to ensure the economic operation of motors.
2) The rated power of the motor should meet the power requirements of the load. It should be based on the load curve which reflects the variation of the load to determine the economic load rate. The appropriate reserve factors are determined by the type and importance of the load. The motor with long-term continuous operation or stable load should make the load rate close to the comprehensive economic load rate. For motors whose year's operation time is larger than 3000 hand the load rate is greater than 60%, the motor of grade 2 or grade 1 efficiency indexes given in Table 2.9 should be the priority for selection.
3) The motor in use should be tested and calculated according to the standard formula and method; determine the motor's actual load rate, efficiency, and power factor. Determine whether it meets the requirements of economic operation; motors that do not meet the requirements of economic operation should be timely modified or replaced.
4) The GB has made detailed provisions for the economic operation and management of the motor. For example, it calls for the establishment of motor running records and has made provisions and a detailed description for inspection and maintenance, instrumentation, power factor compensation, adjustment of operation load, operation monitoring equipment, and motor maintenance.

### 3.3.3 Calculation of Three-Phase Induction Motor Comprehensive Efficiency

#### 3.3.3.1 The Comprehensive Power Loss of the Motor

$$\Delta P_c = \Delta P + K_Q Q \tag{3.25}$$

Substituting Equations (3.14) and (3.18) into the above formula, we obtain

$$\Delta P_c = \Delta P_0 + \beta^2 (\Delta P_N - \Delta P_0) + K_Q [Q_0 + \beta^2 (Q_N - Q_0)] \tag{3.26}$$

where $\Delta P_c$, comprehensive power loss of motor (kW);
$K_Q$, the economic equivalent of the reactive power (kW/kvar).

When the motor is directly connected to the generator bus or reactive compensation bus, $K_Q$ is 0.02–0.04; for the secondary transformer, $K_Q$ is 0.05–0.07; for the third transformer, $K_Q$ is 0.08–0.1. When the grid takes the reactive power compensation, transformer times of motor power should be calculated at the end of compensation.

The above formula, the integrated power loss ($\Delta P_c$), is the total loss including the line loss caused by absorbing reactive power from the grid. It is not only related to the load rate, but related to the reactive power economic equivalent.

#### 3.3.3.2 The Comprehensive Efficiency of the Induction Motor

$$\eta_c = \frac{\beta P_N}{\beta P_N + \Delta P_c} \times 100\% \tag{3.27}$$

where $\eta_c$, the comprehensive efficiency of motor (%).

The so-called integrated efficiency is the operating efficiency that contains line loss.

When the motor load is cyclical load, we can use the weighted average comprehensive efficiency of motor operation.

#### 3.3.3.3 The Weighted Average Comprehensive Efficiency of the Induction Motor Operation

$$\eta_c = \frac{\sum_{i=1}^{n} \eta_{ci} \times t_i}{\sum_{i=1}^{n} t_i} \times 100\%, \tag{3.28}$$

where $\eta_c$, weighted average comprehensive efficiency of motor operation in the investigated period of time (%);
$\eta_{ci}$, the comprehensive efficiency of the motor under load $i$ (%);
$t_i$, the operation time of the motor under load $i$, unit for hours (h);
$n$, the number of motor load changes.

The general assessment of economic operation situation for motors can be directly based on the operation record or load curve to select representative conditions and corresponding run time $t_i$ to calculate the comprehensive efficiency $\eta_c$.

#### 3.3.3.4 The Rated Comprehensive Efficiency of Motor

$$\eta_{cN} = \frac{P_N}{P_N + \Delta P_{cN}} \times 100\%, \tag{3.29}$$

where $\eta_{cN}$, rated comprehensive efficiency of motor (%);
$\Delta P_{cN}$, rated comprehensive power loss (kW).

#### 3.3.3.5 Economic Load Rate of Active Power

$$\beta_P = \sqrt{\frac{\Delta P_0}{\Delta P_N - \Delta P_0}} \times 100\% \tag{3.30}$$

Economic load rate of active power is the load rate of motor when its efficiency is the highest, which does not consider the loss of line caused by reactive power.

#### 3.3.3.6 Comprehensive Economic Load Rate

$$\beta_{cm} = \sqrt{\frac{\Delta P_0 + K_Q Q_0}{\Delta P_N - \Delta P_0 + K_Q \left(Q_N - Q_0\right)}} \times 100\% \tag{3.31}$$

Comprehensive economic load rate is the load rate of motor when its efficiency is the highest, which is the loss of line caused when a reactive power is considered.

In the analysis of whether the three-phase induction motor is in economic operation, the comprehensive efficiency of motor should be calculated first. If it is a cyclic load, the weighted average comprehensive efficiency should also be calculated. Then compared with the rated comprehensive efficiency, determine whether the motor is in economic operation.

### 3.3.4 Judgment of Economic Operation

First judgment method of economic operation: If the comprehensive efficiency of the motor is greater than or equal to the rated comprehensive efficiency, the electrical energy use of the motor is economical; if the comprehensive efficiency of motor is less than the rated comprehensive efficiency but greater than 60% of the rated comprehensive efficiency, the electrical energy use of the motor is basically reasonable; if the motor efficiency is less than 60% of the rated efficiency, the utilization of electrical energy of motor is not economical. The relationship between the motor efficiency and economic operation is shown in Figure 3.2. In the figure, $\eta_c$ represents the comprehensive efficiency and $\eta_{cN}$ represents the rated comprehensive efficiency.

Second judgment method of economic operation: When it is difficult to calculate the comprehensive efficiency of the motor at the scene, the ratio of output power (or current) of motor to the rated input power (or current) can also be used to determine the working status of the motor. Compared with the rated input power (or current), if the input power (or current) is decreased by 15%, it belongs to the economic operation range; if the input power (or current) is decreased within 15–35%, it belongs to basic reasonable range; if the input power (or current) is decreased more than 35%, it belongs to the range of noneconomic operation.

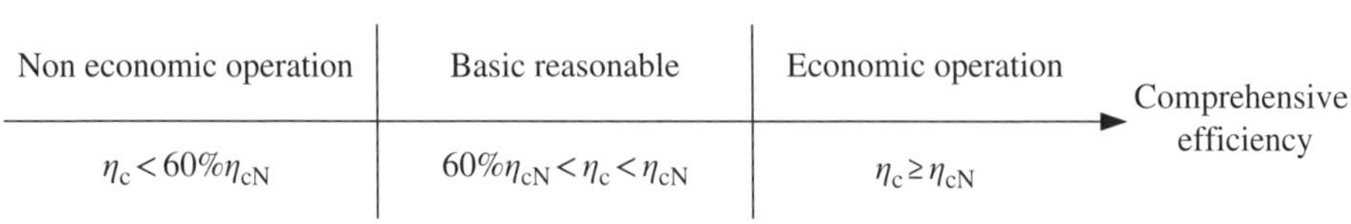


**Figure 3.2** The relationship between the motor efficiency and economic operation.

The essence of energy saving for motor is to improve its comprehensive efficiency, and make it greater than or equal to the rated comprehensive efficiency, namely, motors work in economic operation state. If the load rate can be adjusted to the comprehensive economic load rate $\beta_{cm}$, the motor comprehensive efficiency is the highest, namely the optimal economic operation state of the motor. The comprehensive efficiency of motor should be greater than 60% of the rated comprehensive efficiency, which is the minimum limitation of the economic operation of motor.

In fact, the research of economic operation is centered on the study of "economic operation of three-phase induction motor." Therefore, deep understanding and mastering of the "economical operation of three-phase induction motor" is the premise condition of motor energy conservation research work.

### 3.3.5 The Examples of Economic Operational Analysis

**Example 3.2** In Example 3.1, the load rate is 10, 20, ..., 100%, respectively; calculate the comprehensive power loss and the comprehensive efficiency of the motor. The reactive power economic equivalent is $K_Q = 0.05$. Analyze the economic operation condition of the motor.

**Solution**

In the given example, we know that the rated power is 55 kW, rated efficiency is 0.93, rated power factor is 0.87, rated active power loss is 4.14 kW, rated reactive power is 33.516 kvar, no-load reactive power is 20.02 kvar, and no-load active loss is 1.675 kW.

Rated comprehensive loss of motor

$$\Delta P_{cN} = \Delta P_N + K_Q Q_N = 4.14 + 0.05 \times 33.516 = 5.816\text{kW}$$

Rated comprehensive efficiency of motor

$$\eta_{cN} = \frac{P_N}{P_N + \Delta P_{cN}} \times 100\% = \frac{55}{55 + 5.816} \times 100\% = 90.4\%$$

Considering line loss caused by reactive case, the rated efficiency decreases by 2.6%. When $\beta = 10\%$, the comprehensive power loss of the motor is as follows:

$$\begin{aligned}\Delta P_c &= \Delta P_0 + \beta^2 (\Delta P_N - \Delta P_0) + K_Q \left[ Q_0 + \beta^2 (Q_N - Q_0) \right] \\ &= 1.675 + 0.1^2 (4.14 - 1.675) + 0.05 \times \left[ 20.02 + 0.1^2 (33.516 - 20.02) \right] = 2.707\text{kW}\end{aligned}$$

The comprehensive efficiency of motor

$$\eta_c = \frac{\beta P_N}{\beta P_N + \Delta P_c} \times 100\% = \frac{0.1 \times 55}{0.1 \times 55 + 2.707} \times 100\% = 67.01\%$$

The load rate which is equal to 20, 30, ..., 90 are substituted into formulas (3.26) and (3.27), respectively, and the results are shown in Table 3.1.

As shown in Table 3.1, the comprehensive efficiency is greater than or equal to the rated efficiency, the load rate is only about 90%. If in strict compliance with the

**Table 3.1** Comprehensive loss and comprehensive efficiency under various load rate.

| Load factor (%) | 10 | 20 | 30 | 40 | 50 | 60 | 70 | 80 | 90 | 100 |
|---|---|---|---|---|---|---|---|---|---|---|
| Comprehensive loss (kW) | 2.707 | 2.802 | 2.959 | 3.178 | 3.461 | 3.806 | 4.215 | 4.685 | 5.219 | 5.816 |
| Comprehensive efficiency (%) | 67.01 | 79.70 | 84.79 | 87.38 | 88.82 | 89.66 | 90.13 | 90.38 | 90.46 | 90.44 |

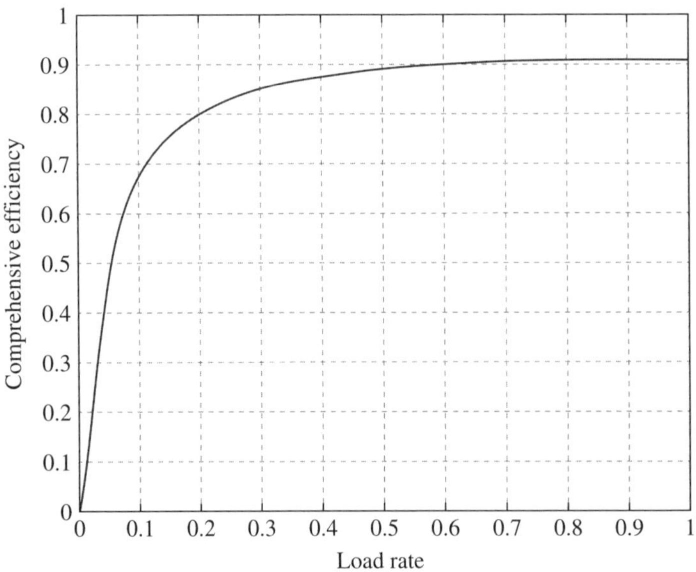


**Figure 3.3** The comprehensive efficiency of Y2-250M-4.

national standard, the economic operation segment of the motor is about 90–100%. In practical work, within 60–100% should be economic operation. Also we note that the "reasonable" conditions are too generous. If in accordance with the conditions of "economic operation of three-phase induction motor," if the comprehensive efficiency is greater than or equal to 60% of the rated comprehensive efficiency, it is considered that the electricity use of the motor is basically reasonable. Then the motors above within 10–60% are "basically reasonable." But if the load rate is less than 30%, the comprehensive efficiency decreases more than 10 percentage points. Therefore, for this motor, load rate of 10–40% is considered uneconomic on energy use; load rate of 40–60% is considered basically reasonable on energy use; load rate of 60–100% is considered economical on energy use.

The comprehensive efficiency curve of the motor is shown in Figure 3.3. Compared with Figure 3.1, it is not hard to see: In considering the loss of line caused by reactive power, flat section (load rate of 50–100%) efficiency has decreased by about 3%; with load rate decreasing, efficiency decreases faster.

**Example 3.3** Known model: Y225M-2 motor's parameters are rated power 45 kW, rated efficiency 0.915, rated power factor 0.89, rated voltage 380 V, rated current 83.9 A, no-load current 18.7 A, no-load losses 1.78 kW, and $K_Q = 0.04$. Requirements:

1) Rated active power loss, rated reactive power, and reactive power of no-load and rated comprehensive efficiency;
2) The comprehensive efficiency when load rate is 60%;
3) The active economic load rate and comprehensive economic load rate.

**Solution**

1) Rated active power loss

$$\Delta P_N = \left(\frac{1}{\eta_N} - 1\right) P_N = \left(\frac{1}{0.915} - 1\right) \times 45 = 4.18\text{kW}$$

Rated reactive power

$$Q_N = \frac{P_N}{\eta_N} \tan\varphi_N = \frac{45}{0.915} \times 0.5123 = 25.195\text{kvar}$$

No-load reactive power

$$Q_0 = \sqrt{3} U_N I_0 = \sqrt{3} \times 380 \times 18.7 = 12.308\text{kvar}$$

Strictly speaking, $\Delta P_0$ should also be considered in no-load reactive power calculations, but $\Delta P_0$ is relatively small and can be ignored.
Rated comprehensive loss

$$\Delta P_{cN} = \Delta P_N + K_Q Q_N = 4.18 + 0.04 \times 25.195 = 5.188\text{kW}$$

Rated comprehensive efficiency

$$\eta_{cN} = \frac{P_N}{P_N + \Delta P_{cN}} \times 100\% = \frac{45}{45 + 5.188} \times 100\% = 89.66\%$$

2) The comprehensive efficiency when load rate is 60%
When $\beta = 60\%$: The comprehensive efficiency of motor

$$\begin{aligned}\Delta P_c &= \Delta P_0 + \beta^2(\Delta P_N - \Delta P_0) + K_Q\left[Q_0 + \beta^2(Q_N - Q_0)\right] \\ &= 1.78 + 0.6^2(4.18 - 1.78) + 0.04 \times \left[12.308 + 0.6^2(25.195 - 12.308)\right] = 3.322\text{kW}\end{aligned}$$

The comprehensive efficiency of motor

$$\begin{aligned}\eta_c &= \frac{\beta P_N}{\beta P_N + \Delta P_c} \times 100\% \\ &= \frac{0.6 \times 45}{0.6 \times 45 + 3.322} \times 100\% = 89.04\%\end{aligned}$$

3) Economic load rate of active power

$$\beta_P = \sqrt{\frac{\Delta P_0}{\Delta P_N - \Delta P_0}} \times 100\%$$
$$= \sqrt{\frac{1.78}{4.18 - 1.78}} \times 100\% = 86.12\%$$

Comprehensive economic load rate

$$\beta_{cm} = \sqrt{\frac{\Delta P_0 + K_Q Q_0}{\Delta P_N - \Delta P_0 + K_Q (Q_N - Q_0)}} \times 100\%$$
$$= \sqrt{\frac{1.78 + 0.04 \times 12.308}{2.4 + 0.04 \times 12.887}} \times 100\% = 88.28\%$$

Considering the line loss caused by reactive power, the maximum efficiency point shifts toward the increasing direction of load rate, the economic operation zone becomes relatively smaller. The larger reactive economic equivalent, the more offset.

## 3.4 Calculation Methods for Energy Saving of the Three-Phase Induction Motor

The energy-saving calculation of the motor is to calculate the difference of motor power consumption before and after renovation, or the difference between the total losses. Before the introduction of motor energy-saving calculation methods, we illustrate two problems.

1) *Energy-Saving Reconstruction*
   Realization of motor energy conservation can have various measures, such as replacing the motor, changing the original operation mode, reducing wear and tear, maintenance, alteration, adjustment of the winding load, adding the speed regulating device, reactive power compensation, and so on. For the sake of convenience, the various energy-saving measures are collectively referred to as the energy-saving reconstruction.
2) *The Premise Condition of Motor Energy-Saving Calculation*
   - Before and after modification for energy saving of motor system, the output power of motor is unchanged. If the load is different, we should make the appropriate conversion.
   - Before and after the motor system energy-saving transformation, power grid voltage is constant.

   Motor energy-saving test and calculation methods under different circumstances are introduced as follows.

### 3.4.1 Using Power to Calculate Energy-saving Amount

This method is mainly used for motor replacement on the spot, testing the input power of the motor before and after the change, then calculating the power-saving rate of the motor before and after the replacement. Using this method should pay attention to that load is unchanged before and after the replacement of the motor, the supply voltage does not change.

#### 3.4.1.1 Active Power Saving

$$\Delta P_p = P_{1q} - P_{1h} \tag{3.32}$$

where $\Delta P_p$, active power-saving value (kW);
$P_{1q}$, the input power of the motor before modification (kW);
$P_{1h}$, the input power of the motor after modification (kW).

#### 3.4.1.2 Reactive Power Saving

$$\Delta Q = Q_{1q} - Q_{1h} \tag{3.33}$$

where $\Delta Q$, conservation value of reactive power (kvar);
$Q_{1q}$, the reactive power of the motor to absorb before modification (kvar);
$Q_{1h}$, the reactive power of the motor to absorb after modification (kvar).

Economic equivalent of reactive power

$$\Delta P_Q = K_Q \times \Delta Q \tag{3.34}$$

where $K_Q$, reactive power equivalent coefficient, accessory methods with the same Equation (3.26).

#### 3.4.1.3 Comprehensive Power Saving

$$\Delta P_c = \Delta P_p + \Delta P_Q \tag{3.35}$$

where $\Delta P_c$, comprehensive power-saving value (kW).
Substituting (3.32) and (3.34) into (3.35)

$$\Delta P_c = \left(P_{1q} - P_{1h}\right) + K_Q\left(Q_{1q} - Q_{1h}\right) \tag{3.36}$$

Equation (3.36) indicates that motor energy saving is to reduce the loss of motor itself; on the other hand it is also to reduce the line loss caused by reactive power, namely reducing the economic equivalent of the reactive power.

#### 3.4.1.4 Calculation of Comprehensive Energy-saving Quantity

$$\Delta W_c = T_{ec} \times \Delta P_c \tag{3.37}$$

where $\Delta W_c$, comprehensive energy-saving quantity of motor (kWh);
$T_{ec}$, operating time (h).

#### 3.4.1.5 Calculation of Comprehensive Power-Saving Rate

$$\Delta E_c = \frac{\left(P_{1q} - P_{1h}\right) + K_Q\left(Q_{1q} - Q_{1h}\right)}{P_{1q} + K_Q \times Q_{1q}} \times 100\% \tag{3.38}$$

where $\Delta E_c$, comprehensive power-saving rate (%).

**Example 3.4** Existing one 45 kW three-phase induction motor drags a constant load. Load variation is that starting needs full load start, after starting the load is reduced to about 40%. The load rate is low; the waste of energy is serious, now to carry out energy-saving technological transformation of the original motor, constant load before and after reconstruction. The measured motor data before transformation: line voltage 380 V, line current 62 A, and power factor 0.45; the measured data after transformation: line voltage 380 V, line current 45 A, and power factor 0.55. Calculate the saving rate of the motor after the transformation, if transformation costs 5000 Yuan, 0.6 Yuan/kW of electricity, calculate the number of days of continuous operation to recover the costs. Coefficient of reactive economic equivalent is 0.04.

**Solution**

The load of this motor is unchanged before and after renovation, namely motor output power $P_2$ unchanged. So the comprehensive energy saving of the motor can directly use the change of the input active power and reactive power to calculate.

Input active power and reactive power before motor transformation

$$P_{1q} = \sqrt{3}U_N I_N \cos\varphi_N = \sqrt{3}\times 380\times 62\times 0.45 = 18.363\text{kW}$$
$$Q_{1q} = \sqrt{3}U_N I_N \sin\varphi_N = \sqrt{3}\times 380\times 62\times 0.893 = 36.44\text{kvar}$$

After transformation

$$P_{1h} = \sqrt{3}U_N I_N \cos\varphi_N = \sqrt{3}\times 380\times 45\times 0.55 = 16.29\text{kW}$$
$$Q_{1h} = \sqrt{3}U_N I_N \sin\varphi_N = \sqrt{3}\times 380\times 45\times 0.835 = 24.735\text{kvar}$$

Active power saving

$$\Delta P_p = 18.363 - 16.29 = 2.073\text{kW}$$

Reactive power saving

$$\Delta Q = 36.44 - 24.735 = 11.705\text{kvar}$$

Comprehensive energy saving

$$\Delta P_c = \Delta P_p + K_Q \times \Delta Q = 2.073 + 0.04\times 11.705 = 2.541\text{kW}$$

Comprehensive energy-saving rate

$$\Delta E_c = \frac{\Delta P_c}{P_{1q} + K_Q \times Q_{1q}}\times 100\% = \frac{2.541}{18.363 + 0.04\times 36.44}\times 100\% = 12.82\%$$

Daily energy-saving money

$$S = 2.541\times 0.6\times 24 = 36.59\text{Yuan}$$

For the investment of 5000 Yuan, 137 days of continuous operation, the cost of technological transformation can be recovered.

### 3.4.2 Comprehensive Efficiency Is Used to Calculate Power-Saving Rate

From (3.38)

$$\Delta E_c = \frac{\left(P_{1q} + K_Q Q_{1q}\right) - \left(P_{1h} + K_Q Q_{1h}\right)}{P_{1q} + K_Q \times Q_{1q}} \times 100\% \tag{3.39}$$

That is

$$\Delta E_c = \frac{P_{1cq} - P_{1ch}}{P_{1cq}} \times 100\% \tag{3.40}$$

where $P_{1cq}$, comprehensive input power before the technical transformation (kW); $P_{1ch}$, comprehensive input power after the technical transformation (kW).

The output power of the motor remain unchanged after the technical transformation, and there

$$\Delta E_c = \frac{P_2/\eta_{1cq} - P_2/\eta_{1ch}}{P_2/\eta_{1cq}} \times 100\% \tag{3.41}$$

where $\eta_{1cq}$, the comprehensive efficiency before technical reconstruction; $\eta_{1ch}$, the comprehensive efficiency after technical reconstruction.

The simplification of formula (3.41) is

$$\Delta E_c = \frac{\eta_{1ch} - \eta_{1cq}}{\eta_{1ch}} \times 100\% \tag{3.42}$$

This method is mainly used in a motor factory for motor energy-saving technological transformation, for example, the dual power transformation introduced later. Test the comprehensive efficiency of motor under one load before modification, then the comprehensive efficiency of the motor under the same load after the transformation. The output power of motor is constant before and after the technical transformation, but the power rate may change, load rate may change too. In the factory, keeping the load and the supply voltage unchanged is relatively easy to realize during the test.

### 3.4.3 Using Accumulated Power to Calculate Power-Saving Rate

In practical problems, the load is often volatile or changing periodically. Therefore, it is difficult to ensure that the load does not change before and after the replacement of the motor. In this case, you can select a relatively long period, such as 1 month. Load in each cycle is relatively stable, or that a particular device's average load in a month is stable. The comparison of electricity consumption of a week or a month before and after replacement of the motor can be used for calculation of power-saving rate of motor. That is

$$\Delta E_c = \frac{\left(W_{pq} + K_Q W_{Qq}\right) - \left(W_{ph} + K_Q W_{Qh}\right)}{W_{pq} + K_Q W_{Qq}} \times 100\% \tag{3.43}$$

where $W_{pq}$, active power of a cycle before replacing the motor (kWh);
$W_{ph}$, active power of a cycle after replacing the motor (kvarh).

Active power can be measured with the watt-hour meter, known as electricity meter. Reactive power can be measured with the reactive watt-hour meter; usually the watt-hour meter and nonpower meter is made into an integration instrument. This method is more intuitive, economic benefit calculation is also very convenient, the method is often applied to motor energy-saving management project.

## 3.5 Comparison and Evaluation Method of Motor Energy-saving Effect

The calculation method of the saving rate introduced above seems to be the evaluation of energy-saving effect of a technological transformation. But this is not a comprehensive evaluation. For instance, an old motor of one device was replaced, and achieving a better energy-saving effect, energy-saving rate reaching 15%. Actually, there are two possibilities: the first one is the original motor which is in serious aging and putting on a new electric motor would have made such a good energy-saving effect; the second is the original motor's matching is not reasonable and putting on a reasonable matching of motor made such a good energy-saving effect. This shows that the energy-saving effect of the motor must determine the reference of energy-saving contrast firstly.

### 3.5.1 Unqualified Old Motor as Reference

In practical work, we usually take the old motor as a reference, measure the input power of the motor before the replacement, change the motor and then measure the input power, compare two values to calculate the power-saving amount. If the old motor has reached retirement age, the motor is unqualified, the test and calculation of the rate of power saving would be meaningless to the evaluation of energy-saving technologies. This case belongs to the update of old motor.

### 3.5.2 Qualified Old Motor as Reference

Although taking the old motor as a reference, but the old motor is qualified, measure the input power of the motor before the replacement, replace the motor, and then measure the input power to calculate the electricity savings. Technical innovation is to solve the irrational matching problem of motor, the energy-saving effect can be used to evaluate the achievements of technological transformation.

### 3.5.3 In Accordance with the National Standard of Motor as Reference

If the purpose of the test is to evaluate a new energy-saving product, or a new energy-saving technology, we should test the original motor first. Make sure that the original motor conforms to the third- level standard of "The small and medium-sized three-phase induction motor efficiency and energy efficiency grade" or not. Then, test the new energy-saving motor and calculate the energy-saving effect. This is the new evaluation method of energy-saving motor.

## 3.6 Discussion Topics of the Chapter

1) For the loss in the three-phase induction motor, which is a constant loss and which is a variable loss?
2) The iron loss is associated with which parameters and variables? What is the method of reducing the loss of iron?
3) Rotor copper loss is related to which parameters and variables? What is the basic method to reduce the copper loss of rotor?
4) The three-phase induction motor running efficiency and power factor is associated with which main parameters, and how about the change rule of curve?
5) Comprehensive power loss, motor comprehensive efficiency, and rated comprehensive, respectively, refer to what?
6) What is the economic equivalent of reactive power and usually based on what values?
7) What is the determining condition for economic operation of the three-phase induction motor?
8) What are the active power energy-saving rate, the reactive power-saving rate, and the comprehensive energy-saving rate?
9) Master the calculation method of comprehensive energy-saving rate and calculate the comprehensive power-saving rate according to actual working condition.

# 4

# The Energy-saving Principle and Method of the Motor Power and Load Match

In terms of motor energy saving, the load factor is very important, so it is necessary to study "lighter load." Most people are of the opinion that "lighter load" brings energy wastage. What is this "lighter load?" Why does the "lighter load" cause energy wastage? In fact, "lighter load" means that the load is unsuitable to the motor. This chapter discusses what the "lighter load" is, and gives the "lighter load" delimitation. Various loss trends will be analyzed under the "lighter load" situation. The chapter also involves the introduction of principles and methods of cooperative power saving in an energy-saving motor. Finally, the energy-saving principle of a dual-power motor will be introduced in the chapter, as well as energy-saving principles and control methods of three common dual-power motors: the double-power series winding motor, double-power winding Y-Δ conversion motor, and extended triangular double-power motor.

## 4.1 Discussion on the "Lighter Load"

### 4.1.1 Boundary of the "Lighter Load"

Generally, the "lighter load" means that the power of the motor is much greater than the load. Then what is the definition of "greater" during lighter load? Strictly, the economic running of motor is the best judgment on "lighter load." For example, in Example 3.2, the load rate of 10–40% is not an economical operation, indicating that this motor (Y2-250M-4) load less than 40% is considered to be the "lighter load." For motors of different powers or different numbers of poles, critical point of noneconomic running is different. Critical points of noneconomic running may appear when the load is 40%, 45%, or at a point of 50%. This is different from "lighter load."

In order to discuss this problem, we analyze the characteristic of the three-phase induction motor efficiency and power factor curve again. The efficiency and power factor curves of the three-phase induction motor (Y280S-6, 45kW/6) are shown in Figure 4.1.

The efficiency of the three-phase induction motor in the graph is changed with output power. The efficiency curve can be roughly divided into three sections: a-o section, a-b section, and b-c section. Within the a-o section, the efficiency curve falls sharply, or mean efficiency rapidly increases with the increase of load; the a-b section efficiency changes obviously; the b-c section efficiency changes gently, close to the rated value. Point a, the turning point of the efficiency, is called the critical point. This point is called

*Energy-saving Principles and Technologies for Induction Motors*, First Edition. Wenzhong Ma and Lianping Bai.

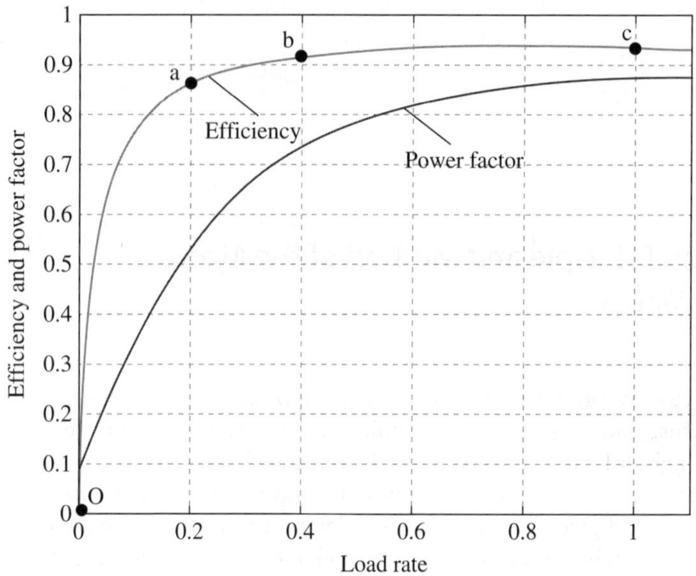


**Figure 4.1** Y280S-6 efficiency and power factor curves.

the critical load rate $\beta_a$ corresponding to point a. If $\beta > \beta_a$, efficiency slightly changes. This is due to the relation between the variable loss of motor and constant loss. If $\beta < \beta_a$, there is a sharp drop in efficiency. When $\beta$ is greater than $\beta_a$, despite the high efficiency, power factor is lower. In the figure, it is only 0.53. At this point, the comprehensive efficiency is 10% lower than the rated efficiency.

Usually, a three-phase induction motor of several kilowatts is used. For this, we focus on another example. The three-phase induction motor (Y100L-4) is 3 kW/4 pole, and its efficiency and power factor curve is shown in Figure 4.2.

In Figure 4.2 the efficiency curve in general can be divided into three sections: section a-o, section a-b, and section b-c. In contrast with Figure 4.1, the difference is that the efficiency and power factor with load rate rise slower. $\beta_a$ moves from 0.2 to 0.3, and $\beta_b$ moves from 0.4 to 0.5. Research shows several kilowatts or kilowatts of efficiency and power factor of a three-phase induction motor curve and what is shown in Figures 4.1 and 4.2 are approximately the same. The larger the power, the faster the rise in efficiency and power factor.

In a word, the definition of *t* "lighter load" of motors should include the efficiency and power factor. For different power and number of poles of the three-phase induction motor, efficiency and power factor curves are different. Strictly, the proportions of "lighter load" are not clear. But, in general, for the average Y series motor (Y, Y2, Y3, and YX) more than 1.5 kW, if $\beta \leq 50\%$, it can be called "lighter load."

Then what is the appropriate proportion of "horse" and "car?" From the perspective of the power match, load rate from 70–80% is proper. When motor efficiency is the highest, the power factor is high. In case of load volatility, load rate from 70–80% will reserve some allowance for motor power. Although the cost increased, it can improve the reliability of the motor drive system.

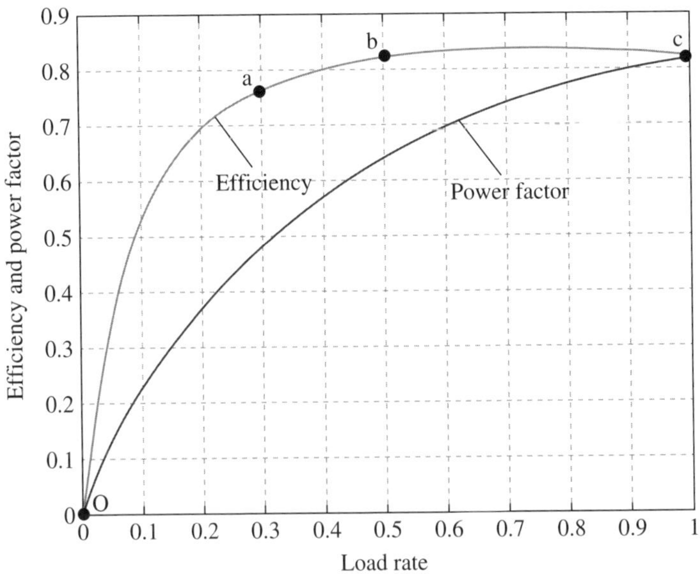


**Figure 4.2** Y100L-4 efficiency and power factor curves.

### 4.1.2 Analysis of the Lighter Load Loss

From the efficiency and power factor curves, we come to know that when motors keep "lighter load" both the efficiency and the power factor are low. This is the reason why "lighter load" leads to energy wastage. Strictly, the loss issue is the main reason of energy wastage. For a certain load, if the original working situation is "lighter load" and a small change occurs in the motor power, how will the five kinds of loss change? Analyzing the loss when the motor is under "lighter load" with the help of an example:

**Example 4.1** A 14-type beam pumping unit, whose average load is 11.25 kW. The original motor (Y315M-8) is 75 kW, and it is in serious "lighter load" state. The average load rate is 15%. Then we design a high starting torque motor, and it solves the problem of starting difficulty. The Y250M-8 motor powers down to 30 kW, and the average load rate increased to 37.5%. There is no change whether the motor load is replaced or not. After testing and calculation, we obtain the various loss changes before and after the two electric motors, as shown in Table 4.1.

From Table 4.1, for the same load, two electric motors of various losses vary a lot. Respectively, the percentage of all loss reduced: iron loss dropped by 75%; stator copper loss dropped by 46%; rotor copper loss dropped by 72%; mechanical consumption

**Table 4.1** All kinds of loss of two electric motors (kW).

| Motor model | Input | Iron loss | Stator copper loss | Rotor copper loss | Mechanical loss | Stray loss |
|---|---|---|---|---|---|---|
| Y315M-8 | 14.670 | 1.850 | 0.560 | 0.230 | 0.587 | 0.193 |
| Y250M-8 | 12.458 | 0.462 | 0.303 | 0.065 | 0.310 | 0.068 |

**Table 4.2** Y315M-8/75kW the change of motor various losses (kW).

| Load factor (%) | Input | Iron loss | Stator copper loss | Rotor copper loss | Mechanical loss | Stray loss |
|---|---|---|---|---|---|---|
| 100 | 81.062 | 1.865 | 1.740 | 0.950 | 0.587 | 0.920 |
| 15 | 14.670 | 1.850 | 0.560 | 0.230 | 0.587 | 0.193 |

dropped by 47%; stray loss dropped by 65%; and iron loss goes down obviously. For motor Y315M-8/75kW, the total loss was 3.42 kW. Among them, the iron loss occupied 54.1%. This shows that the energy wastage in "lighter load" is mainly because of the iron loss, followed by the stator copper loss and mechanical loss.

Then analyze the loss changes of "lighter load." For motor Y315M-8/75kW, all kinds of losses from rated load to half a quarter are shown in Table 4.2.

From Table 4.2, when the load changes from rated load to lighter load, the iron loss and mechanical loss of the motor basically remain unchanged. Copper loss and stray loss are much smaller.

When Y315M-8/75kW motor is running at a rated load, the efficiency is 92.5% and the power factor 0.817. When the average load rate is 15%, efficiency is 76.7%, power factor is 0.435, and the total loss is 3.420 kW.

The efficiency of Y250M-8/30kW motor running at rated load is 91.6% and the power factor is 0.810. When the load is 37.5%, efficiency is 90.3% and power factor is 0.577. The total loss is 1.208 kW. From the data above, we can conclude that the load rate is increased when the 75 kW motor is replaced by a 30 kW motor and the total loss drops from 3.42 to 1.208 kW. The motor input power dropped from 14.670 to 12.458 kW. Active energy-saving rate is 15.1%.

In this example, the load is a special kind of load. The average load rate is not able to be increased to 50%, not to mention 60%. The load makes the motor still remain in the "lighter load" state. Thus beam pumping unit motor energy conservation has yet to be deeply studied.

## 4.2 Energy-saving Principle of Power Matching

In Section 4.1, we conducted an analysis on "lighter load." The main reason why "lighter load" causes energy wastage is the iron loss, followed by the stator copper loss and mechanical loss. The key to solve the "lighter load" problem is to achieve the rational matching power and load power. In the production practice, a lot of motor power match is not reasonable. So, in the motor energy-saving process, solving the "lighter load" problem is a main aspect of electric motor energy conservation. Following is the power matching principle of energy conservation, and methods to solve the "lighter load" problem.

### 4.2.1 The Power Matching Principle of Energy Conservation

From the perspective of power matching, to achieve motor energy saving, actually it satisfies the "three-phase induction motor and economic operation" conditions.

That means if the comprehensive efficiency of a motor is greater than or equal to the rated comprehensive efficiency, the motor is in an economic operation state. Otherwise, the motor is not economic to use. Determining whether the motor operates economically or not, is to look at the comprehensive efficiency when the motor is running. According to comprehensive analysis in Chapter 3, calculation of motor efficiency is very complex. The conditional judgment mentioned is not clear. So the economic operation of motor in judgment encountered more difficulties. Through the analysis and calculation of the comprehensive efficiency of the motor, we see that if the power factor is reduced slightly, the comprehensive efficiency is close to the rated comprehensive efficiency. So you can replace the comprehensive efficiency with an active efficiency motor, and directly use the load factor to determine whether a motor is in economic operation.

Through following an electric motor (Y200L-6/22kW) parameter test and calculation, the above conclusion can be verified. On the motor test bench, gradually adjust the load factor of the motor, and test the motor efficiency and power factor. Start to increase the load rate with a 5% step to 40%, and with 10% step increasing to 100%. The measured efficiency ($\eta$) and power factor are shown in Table 4.3. The comprehensive efficiency of calculation and analysis will be given.

The rated active loss:

$$\Delta P_{\mathrm{N}} = P_{\mathrm{N}}\left(\frac{1}{\eta_{\mathrm{N}}}-1\right)=22\times(1.11-1)=2.42\mathrm{kW}$$

The rated reactive power:

$$Q_{\mathrm{N}} = \frac{P_{\mathrm{N}}}{\eta_{\mathrm{N}}}\tan\varphi_{\mathrm{N}} = \frac{22\times 0.63}{0.901}=15.383\mathrm{kvar}$$

The rated comprehensive loss:

$$\Delta P_{\mathrm{cN}} = \Delta P_{\mathrm{N}} + K_{\mathrm{Q}}Q_{\mathrm{N}} = 2.42+0.02\times 15.383 = 2.728\mathrm{kW}$$

When testing, there is concentrated compensation before the third transformation, so $K_{\mathrm{Q}}$ is 0.02.

The comprehensive efficiency rating:

$$\eta_{\mathrm{cN}} = \frac{P_{\mathrm{N}}}{P_{\mathrm{N}}+\Delta P_{\mathrm{cN}}}\times 100\% = \frac{22}{22+2.728}\times 100\% = 88.9\%$$

**Table 4.3** The test parameters of the Y200L-6/22kW motor.

| | | | | | | | | | | | | | | | |
|---|---|---|---|---|---|---|---|---|---|---|---|---|---|---|---|
| $\beta$ | 0.01 | 0.05 | 0.10 | 0.15 | 0.20 | 0.25 | 0.30 | 0.35 | 0.40 | 0.50 | 0.60 | 0.70 | 0.80 | 0.90 | 1.00 |
| $\eta$ | 0.256 | 0.623 | 0.757 | 0.815 | 0.846 | 0.865 | 0.878 | 0.887 | 0.893 | 0.901 | 0.904 | 0.905 | 0.905 | 0.903 | 0.901 |
| Power factor | 0.075 | 0.153 | 0.246 | 0.332 | 0.411 | 0.480 | 0.540 | 0.592 | 0.636 | 0.705 | 0.754 | 0.789 | 0.805 | 0.832 | 0.846 |

In the table above if $\beta = 0.6$, we can get the active loss:

$$\Delta P_{\beta} = P_2\left(\frac{1}{\eta_{\beta}} - 1\right) = 13.2 \times (1.106 - 1) = 1.40\text{kW}$$

The reactive power:

$$Q_{\beta} = \frac{P_{\beta}}{\eta_{\beta}} \tan\varphi_{\beta} = \frac{13.2 \times 0.871}{0.904} = 12.718\text{kvar}$$

The comprehensive loss:

$$\Delta P_{\text{c}} = \Delta P_{\beta} + K_{\text{Q}} Q_{\beta} = 1.40 + 0.02 \times 12.718 = 1.654\text{kW}$$

The comprehensive efficiency:

$$\eta_{\text{c}} = \frac{P_2}{P_2 + \Delta P_{\text{c}}} \times 100\% = \frac{13.2}{13.2 + 1.654} \times 100\% = 88.9\%$$

In the table above, if $\beta = 0.3$,

$$\Delta P_{\beta} = P_2\left(\frac{1}{\eta_{\beta}} - 1\right) = 6.6 \times (1.139 - 1) = 0.917\text{kW}$$

$$Q_{\beta} = \frac{P_{\beta}}{\eta_{\beta}} \tan\varphi_{\beta} = \frac{6.6 \times 1.558}{0.878} = 11.712\text{kvar}$$

$$\Delta P_{\text{c}} = \Delta P_{\beta} + K_{\text{Q}} Q_{\beta} = 0.917 + 0.02 \times 11.712 = 1.151\text{kW}$$

$$\eta_{\text{c}} = \frac{P_2}{P_2 + \Delta P_{\text{c}}} \times 100\% = \frac{6.6}{6.6 + 1.151} \times 100\% = 85.15\%$$

It can be seen from the calculation above that if $\beta = 0.6$, the power factor drops below the rated power factor by 0.10. At this point the overall efficiency equals the rated overall efficiency. Motor conforms to economic operating conditions. When $\beta = 0.3$, the power factor drops below the rated power factor by 0.306. The comprehensive efficiency (85.15%) is less than the rated comprehensive efficiency. Motor does not correspond to the requirements of the economic operation. But it is essential for the rational use of energy. So if load rate is greater than 60%, the motor meets the economic operating conditions. Then the following conclusions are obtained.

*Commonly Y series (Y, Y2, Y3, and YX) motors above 1.5 kW and from the angle of power matching, it is necessary to try to achieve more than 60% load factor to achieve economic operation of the motor. This is the power matching energy-saving principle of an induction motor.*

How to realize the load factor more than 60%? We need to further study the actual situation of motor drive system. There are two common situations: one is the original motor drive system, and the other is to design a new motor drive system. If the original motor drive system keeps in the "lighter load" state for a long time, it is necessary to replacc the motor. If you want to design a new motor drive system, a reasonable power matching motor according to the load situation is needed. Therefore, the reasonable choice of the motor power is particularly important.

### 4.2.2 Motor Selection Steps

The basic steps of motor selection: type selection → voltage selection → speed selection → torque selection → power selection.

1) Type Selection. The load characteristics of the driving machinery should be fully understood before motor selection. When the load does not need special requirements on starting, braking, and speed, squirrel cage induction motors should be chosen first. Otherwise, the choice of motor shall meet the corresponding torque and maximum torque requirements. The selected motor should be able to reasonably match the speed adjustment mode. In Table 2.16, the main types, characteristics, and application of three-phase induction motors are listed. Type selection needs some detailed understandings of characteristics and applications of various types of motors. It should take account of the work environment, whether the motor is working in an inflammable, explosive, dust pollution, corrosive gas, high temperature, high altitude, high humidity, water spray, and scuba diving environment. According to the environmental limitation, choose the corresponding protective type, and the insulation of the motor protection grade level. If there are special requirements for motor noise, we should choose a low noisy motor.
2) Voltage Selection. The working voltage of the motor should be adapted to the power voltage supplied. Common supply voltage is 380 V, whereas 660 V power is supplied on special occasions. High voltage motor works on 1.14, 3, 6, and 10 kV power voltage. A motor whose rated power is greater than 200 kW should be preferred for a high voltage motor. Motor running in speed regulation mode should be chosen at rated voltage levels.
3) Speed Selection. The speed of the motor should be the same as the load speed or similar. If the speed needs to be decreased (or increased), with the requirements of transmission satisfied, mechanical drive series should be reduced as much as possible. Choose a reasonable way of speed adjustment and electric motors according to the speed range, efficiency, counter rotating torque, and the influence of the long-term economic benefits and other factors.
4) Torque Selection. The motor should meet the locked-rotor torque load and peak torque demands. We should choose the corresponding special motor (such as YR or YH series) and torque calibration in cases of frequent starting, impact load, and high starting torque load.

In addition, when the running time of a year is more than 3000 h and load rate is more than 60%, we should first select the level of motor (grade 2 or grade 1) mentioned in Table 2.9.

### 4.2.3 The Selection of the Motor Rated Power

Correct selection of the motor rated power is vital to the improvement of the economic operation level of motor system, energy saving, and economic benefit. Motor power options should be properly meeting the requirements of the situation of production machinery. If the power is low, normal operation of production machinery cannot be guaranteed, and this might even burn the motor. If the power is too large, it is an uneconomical operation and brings lighter load.

#### 4.2.3.1 Requirements of Power Selection

1) Motor rated power should meet the requirement of power load. Torque can satisfy the load torque and torque requirements.
2) The heating should be close to the allowable working temperature in operation, but not too much.
3) Motors need overload capacity, in order to ensure the overload operation of the short time.
4) Depending on the load curve reflects the load changes, to determine the economic load rate.

#### 4.2.3.2 Steps of Power Selection

1) Calculation of load power or load torque.
2) According to the torque or power, preselect motor rated power.
3) Check motor of fever, starting capability, and overload capacity.

The following example shows the process of motor power selection.

**Example 4.2** There is a centrifugal pump and its voltage is 380 V. The pump speed is 1450 r/min, pumps 9.5 $m^3$/min, total 4 m, and pump efficiency is 80%. Internal maximum temperature should not exceed 30 °C. Try to select motor type and power.

**Solution**

1) Type Select: It can be seen from Table 2.16 that we need three-phase squirrel cage induction motors; hence, choose Y2 series because of its high efficiency.
2) Voltage Selection: Pumping power supplied is 380 V, so the motor voltage must be 380 V.
3) Speed Selection: Speed of pump close to 1500 r/min, you can choose a 4-pole motor.
4) Torque Selection: The request of the pump is not high on the starting torque.
5) Power Selection:

Step 1. Calculating power load
The pump power can be calculated by the formula

$$P_{sb} = k_p \frac{QH}{6.12\eta_p} \tag{4.1}$$

where $P_{sb}$, the shaft power of pump (kW);
$k_p$, pump power margin coefficient. From 1.1 to 1.2;
$Q$, pump discharge ($m^3$/min);

$H$, total head (m);
$\eta_p$, pump efficiency.

Substituting the data

$$P_{sb} = k_p \frac{QH}{6.12\eta_p} = 1.2 \times \frac{9.5 \times 4}{6.12 \times 0.8} = 9.3\text{kW}$$

Step 2. The primary motor power according to the results of the calculation.
It is seen from Table 2.13, exceeding 9.3 kW is 11 kW, so it can only choose 11 kW. If the pump works in rated condition, the motor load rate is 84.7%. Load rate is high, and there is 15% margin.

Look up Table 2.13, optional type of motor is Y2-160M-4/11kW. The speed of the motor is 1460 r/min, a little higher than the speed of the pump; rated current is 22.2 A and rated efficiency is 88%. Rated power factor is 0.84 and starting current ratio is 7. The overload capacity is 2.3 and starting torque is 2.2. The minimum torque is 1.4.

Step 3. Check the motor starting power of fever, and overload capacity

1) Motor Heat Checking: The above condition is a continuous work in the pump system, using the equivalent power method to verify fever.

   When the environment temperature $\theta_0 = 30\,°\text{C}$, we modify the rated power of motor. $\alpha = 0.5$ (the ratio of constant loss and variable loss), and the steady-state temperature is 75 °C under rated load. Calculation formula checking power is as follows

$$P_{JY} = P_{sb}\sqrt{1 + \frac{40 - \theta_0}{\tau_{we}}(\alpha + 1)} \tag{4.2}$$

   where $P_{JY}$, motor heat check power (kW);
   $\tau_{we}$, rated load steady state temperature (°C).

   Substituting the known data

$$P_{JY} = 9.3\sqrt{1 + \frac{40 - 30}{75}(0.5 + 1)} = 10.2\text{kW}$$

   It can be concluded that check power is less than the selected motor. In addition, Y2 series motor insulation is F level and the highest temperature is 150 °C. So it meets the load requirements of motor temperature rise.

2) Motor Starting Capability Check: Starting torque of the selected motor is 2.2 and the minimum torque is 1.4. According to the national standards (GB12497-1995), centrifugal pump load torque is the torque at a rated load of 0.4–0.6 times. Motor starting torque and the minimum torque are greater than the load torque. Moreover, there is 15% margin. So the starting capacity of the selected motor meets the load requirements.
3) Check the Overload Ability of Motor: According to the national standards (GB12497-1995), centrifugal pump maximum load torque is 1.5 times the rated load torque. Overload capacity of the motor selected is 2.3 times. With 15% of the margin, the overload capacity is enough.

**Example 4.3** A workshop has a bridge-type hook crane. The maximum lifting load is 5 tons (including hook quality and cable quality). Lifting and lowering speed is 12 m/min, braking torque and hoisting torque are same. The speed reduction ratio between the electric motor and the drum is 75. Roller radius is 0.2 m. The overall efficiency of crane hoisting mechanism is 80%. The working process of the crane lifting weight: lifting 30 s, stopping 2 s, walking 60 s, stopping 2 s, lowering the weight 30 s, stopping 2 s, and return to walking 60 s. Workshop power supplied is 380 V. In the workshop environment, the highest temperature is not more than 35 °C. Try to choose the type of crane hoist motor and power.

**Solution**

1) Type Selection: Crane equipment requires a high starting torque and large overload capacity; we should choose motors of YZ or YZR series. Because starting switching equipment of motors of YZR series is complex, we have better chosen a motor of YZ series.
2) Voltage Selection: According to the workshop power supply voltage, the motor voltage is 380 V.
3) Speed Selection: Relationship between motor speed and line speed.

$$v_1 = \frac{2\pi Rn}{i} \tag{4.3}$$

where $v_1$, speed of crane lifting (m/min);
$n$, motor speed (r/min);
$R$, the mean radius of the drum (m);
$i$, gear transmission ratio.

$$n = \frac{iv_1}{2\pi R} = \frac{75\times 12}{2\times 3.14\times 0.2} = 716.6\,\text{r}/\text{min}$$

The speed close to 716.6 r/min is 750 r/min, so we choose an 8-pole motor.
4) Torque Selection: Motor torque can vary because of the power, so make sure about the power, and then check the torque.
5) Power Selection:

Step 1. Calculation of load power
Power of crane is calculated by as follows:

$$P_{qz} = \frac{W_1 v_1}{6120\eta_1} \tag{4.4}$$

where $P_{qz}$, crane load power (kW);
$W_1$, the weight of the object, hook, and rope (kg);
$v_1$, lifting speed (m/min);
$\eta_1$, crane efficiency.

Substituting the data:

$$P_{qz} = \frac{W_1 v_1}{6120\eta_1} = \frac{5\times 10^3\times 12}{6120\times 0.8} = 12.26\,\text{kW}$$

Step 2. Proselecting motor power according to the results of the calculation.
It is seen from Table 2.13 that 15 kW is greater than 12.26 kW, so we choose 15 kW. When the crane is full load, the load rate is 82% and the load rate is high. Therefore, there is 18% margin.

Refer to the product manual of the crane motor and select the model YZ200L-8/15kW. Duty for S3, rated load duration rate is 40% and the motor's rated speed is 710 r/min. The rated current is 33.1 A. Rated efficiency is 86.2% and rated power factor is 0.80. Starting current ratio is 6.1 and the overload capacity is 2.5 times rated load. The starting torque ratio is 2.2.
Motor torque check:

Rated torque

$$T_N = 9550\frac{P_N}{n_N} = 9550 \times \frac{15}{710} = 202\text{Nm}$$

Load torque of roller

$$T_\text{L} = W_1 gR = 5 \times 1000 \times 9.8 \times 0.2 = 9800\text{Nm}$$

Drive torque of roller

$$T'_\text{N} = iT_\text{N} = 75 \times 202 = 15{,}150\text{Nm}$$

$$T'_\text{N} > T_\text{L}$$

Step 3. Check the motor starting power of fever, and overload capacity

1) Motor Heat Checking:
   Practical duty cycle

$$FC_\text{L} = \frac{30+30}{30+2+60+2+30+2+60} \times 100\% = 32.3\%$$

   Motor rated load sustained rate is 40%. Motor heating power for checking is as follows:

$$P_\text{JL} = P_\text{L}\sqrt{\frac{FC_\text{L}}{FC_\text{N}}} = 12.26\sqrt{\frac{32.3}{40}} = 9.9\text{kW}$$

$$P_\text{N} > P_\text{JL}$$

2) Motor Starting Capability Check: The starting torque ratio of the motor selected is 2.2. According to the national standards (GB12497-1995), crane starting torque is 1.6–2.0 times rated load torque. The starting torque of the motor is greater than the load, with 18% margin. Overload capacity satisfies the requirement of load.
3) Check the Overload Ability of Motor: The overload capacity of motor selected is 2.5. According to the national standards (GB12497-1995), crane peak torque is 1.8–2.2 times of rated load torque. Overload capacity is greater than the load demand of overload, with 18% margin. So the overload ability is enough.

Type selection and power selection are the key to motor selection. The two examples above mainly illustrate the basic method of motor selection. The mechanical load in practice is varied, and the calculation of load power is very complex. Calculation of complex mechanical load power needs access to relevant mechanical design manual. For deeper questions on motor power selection, please refer to the national standards (GBl2497-1995), where detailed provisions are provided.

In short, the motor type selection and power selection are the key to solve the "lighter load" problem. Calculated results may not be entirely consistent with the actual situation. So it needs investigation and verification. Under permitting conditions, we should test the original (or similar) motor drive system when motors keep running. By comparing the theoretical calculation and the actual situation, we can make better motor type selection and power selection.

## 4.3 Double Power Induction Motors and Energy-saving Principle

### 4.3.1 Double-Power Induction Motors

As the motor drive system is not content with economic operation to technical transformation, we do not tend to replace the motor, but carry out technical reform for the original motors, because the cost of technical reform is much less than the cost of purchasing new motors. When we carry out technical reform for the original motor, using the method of double-power combination can achieve high-energy efficiency to motors. The following illustrates energy-saving principle of the double-power motor.

There is a motor drive system with a large fluctuation of loads torque, about the average load of 15 kW, and the original motor is 37 kW. The motor is always in "lighter load," so waste of energy is serious. Now we redesign the stator winding, which can be designed to complement double power: one is 37 kW; the other is 22 kW. When the load is greater than 22 kW, we can put in a winding operation of 37 kW, so the motor load rate is greater than 60%; when the load is about 15 kW, we can put in a winding operation of 22 kW, so the motor load rate is still greater than 60%. Although the loads torque has a large fluctuation, the motor still has an economical load rate. So it is very good to solve "lighter load." Motor efficiency and power factor have a larger improvement, meeting the motor economic operation conditions.

According to the range of loads change, the stator windings of the motor are divided into two: high-power winding and low-power winding. Detecting the motor current in the process of the motor running, determining the size of loads according to the relation between current and power, making sure to put into the appropriate winding, makes the power of motors and load match. Although the loads torque has a large change, the motor still has a high load rate. *This is the energy-saving principle of the double-power induction motors.*

Actually, aimed at the actual situation of the large change of load, the design of the new motor can also adopt this method.

And then, if the motors are designed as three-power, is it better for economic operation? The 37 kW motor is designed to three-power winding: 37 kW, 22 kW, and 15 kW, respectively. Unless the load power is reduced to less than 11 kW, we do not design the power of 15 kW. Because the main body of motor remains unchanged, and the iron loss does not drop too much, the low-power winding energy saving is not of great significance.

**Something needs to be pointed out:** The double-power windings induction motors do not run in parallel mode. When the double-power windings switch, the parallel windings are used in alternation. That is to say that only one winding has electricity at a time, leading to low utilization of winding material; meanwhile copper loss cannot be effectively reduced, and energy efficiency is not satisfactory. So, double-power motors do not adopt in parallel. The windings of dual-power motors usually adopt these three kinds of designing methods as follows.

1) The Series of Power-Winding (for Short Series Winding). Stator winding is designed to two windings: one is the main winding; the other is assistant winding. The main winding is high-power winding, and two windings in series is low-power winding. According to characteristics of loads, running motor is switched.
2) Y-Δ Switching of Double-Power Windings. We all know that three-phase windings of motors are connected in Δ-windings to output high power, and in star-windings to output low power. According to loads, the connection pattern of windings is switched in operation of the motor.
3) Extended Δ Connected Dual-Power Windings. Each phase winding of motors has a tap. Do not trip the tap at the beginning, connecting three-phase windings Δ makes the motor output high power; or connecting part of three-phase windings and taps into small Δ, then the other taps connect extended sides of Δ, so it is called extended Δ connected windings. Extended Δ connected windings are low-power windings. According to loads, the windings connection style is switched in operation of the motor.

The following part introduces energy-saving principle, energy efficiency, and the control method of switching windings for the series winding.

### 4.3.2 Energy-saving Principle of the Double-Power Motors

Based on principle of designing motors, if the excitation current can be reduced, it not only reduces the copper loss and iron loss of the stator, but also improves the power factor of motors. Reducing the excitation current is an effective way to improve the power factor of motors. The power factor of motors is as follows:

$$\cos\varphi = \frac{I_{1p}^{*}}{\sqrt{I_{1p}^{*2} + I_{1q}^{*2}}} \tag{4.5}$$

where $I_{1p}^{*}$, the stator active current unit value;
$I_{1q}^{*}$, the stator reactive current unit value;

$$I_{1q}^{*} = I_{m}^{*} + I_{x}^{*} \tag{4.6}$$

where $I_{m}^{*}$, the stator excitation current unit value at load rating;
$I_{x}^{*}$, the rotor reactive current component unit value at load rating;

of which $I_m^*$ is the main portion, it is above 60% of the total. So we focus on which parameters contribute to $I_m^*$. As known from the principle of motor

$$I_m = \frac{2pF_0}{0.9m_1N_1K_{dp1}} \tag{4.7}$$

where $m_1$, the number of phase of stator winding;
$p$, the pole pairs;
$K_{dp1}$, the factor of fundamental wave of the stator winding;
$N_1$, the number of turns of each phase of the serial windings of the stator;
$F_0$, the magneto-motive force of each pole.

As can be seen from the above three formulas, the most direct method to improve the power factor $\cos\varphi$ is to increase the number of turns of each phase of the serial windings of the stator. But increasing the number of turns can result in new problems; it will reduce the starting torque of motors. This is to say, that $N_1$ increases as the starting torque decreases, and it is contradictory. In order to solve this problem, we can divide stator winding into two portions. A part is the main windings, which is the multicircuit winding, and the other part is additional winding, which is the minor-circuit winding. Two windings can cascade together. Under starting up or heavy loads, the main winding is activated. Under the lighter load, the additional winding in series turns into the low-power winding. Based on the original motor, technical reconstruction is carried out; wire diameter of windings and number of windings can be only properly designed by ascertaining the tank size.

For example, the original motor is 37 kW. The stator winding is designed into two windings. The power of the main winding is 37 kW as a small winding in series, which the power is 22 kW after the small winding in series. When the load is greater than 22 kW, the winding of 37 kW should be switched into operation, and the load rate is very high. When loads are small, the high-power winding should be switched into the low-power winding (that puts into the winding of 22 kW) to operate. Then the load rate is still very high. The excitation current of motors in itself reduces, iron loss reduces, and the copper loss of the stator reduces. It has greatly enhanced efficiency and power factor for motors, so lighter-load is well resolved. This is the energy-saving principle of the double-power motors.

### 4.3.3 Analysis of the Energy-saving Effect of Winding in Series

In order to analyze the energy-saving effect of the dual-power winding in series, as a result of the comparison test of two motors, the result is obvious. First choose a motor in line with national standards (hereafter referred to standard motors), that model is Y280M-6/55 kW. In test experimental bed of motors, gradually adjust the load rate of motors, test efficiency, and power factor of motors, and document test results in Table 4.4. For the design, manufacture good dual-power winding motor (hereafter referred to the energy-saving motor) to do the testing, the energy-saving motor frame size is the same as the standard motor that is Y280M-6, among the high-power winding is 55 kW and the small-power winding is 30 kW. Testing results of efficiency and the

**Table 4.4** Efficiency and power factor of two kinds of motors.

| | Y280M-6 (the standard motor 55 kW) | | Y280M-6 (the energy-saving motor 30/55 kW) | |
|---|---|---|---|---|
| The motor loading rate $\beta$ (%) | Efficiency $\eta$ | The power factor cos $\Phi$ | Efficiency $\eta$ | The power factor cos $\Phi$ |
| 1.82 | 0.4572 | 0.1033 | 0.6632 | 0.1803 |
| 3.64 | 0.6237 | 0.1505 | 0.7908 | 0.2931 |
| 7.27 | 0.7620 | 0.2415 | 0.8737 | 0.4810 |
| 10.91 | 0.8224 | 0.3261 | 0.9041 | 0.6151 |
| 14.55 | 0.8561 | 0.4032 | 0.9191 | 0.7063 |
| 18.18 | 0.8772 | 0.4716 | 0.9276 | 0.7666 |
| 21.82 | 0.8916 | 0.5314 | 0.9326 | 0.8076 |
| 25.45 | 0.9019 | 0.5832 | 0.9355 | 0.8356 |
| 29.09 | 0.9096 | 0.6278 | 0.9372 | 0.8550 |
| 32.73 | 0.9155 | 0.6659 | 0.9380 | 0.8686 |
| 36.36 | 0.9200 | 0.6983 | 0.9381 | 0.8780 |
| 40.00 | 0.9236 | 0.7261 | 0.9377 | 0.8844 |
| 43.64 | 0.9264 | 0.7495 | 0.9370 | 0.8885 |
| 47.27 | 0.9287 | 0.7697 | 0.9359 | 0.8909 |
| 50.91 | 0.9305 | 0.7869 | 0.9345 | 0.8919 |
| 54.55 | 0.9331 | 0.8145 | 0.9274 | 0.8174 |
| 65.45 | 0.9345 | 0.8349 | 0.9285 | 0.8375 |
| 72.72 | 0.9354 | 0.8501 | 0.9288 | 0.8526 |
| 80.00 | 0.9356 | 0.8615 | 0.9285 | 0.8638 |
| 87.27 | 0.9354 | 0.8699 | 0.9278 | 0.8722 |
| 94.55 | 0.9348 | 0.8760 | 0.9267 | 0.8783 |
| 100 | 0.9342 | 0.8795 | 0.9256 | 0.8816 |

power factors are listed in Table 4.4. Prepare the curve of efficiency and the power factors of two kinds of motors shown in Figures 4.3 and 4.4. From Figure 4.3, you can see that the efficiency of the energy-saving motor is slightly higher than the standard motor's in lighter load. From the data of Table 4.4, you can see that the efficiency of the energy-saving motor after 30 kW is less than one percentage point below its figure for the standard motor's. As can be seen from Figure 4.4, the power factor of the energy-saving motor before the power switch point is a lot higher than the standard motor's, but the power factors of two kinds of motors are fairly similar after the power switch point. From the comparison from the curve of efficiency and the power factor, you can see that energy-saving effect of the dual-power winding in series motor is very obvious.

The following is theory analysis for the dual-power winding in series of motors. Theoretically the energy-saving effect should be the comparison of two curves and two

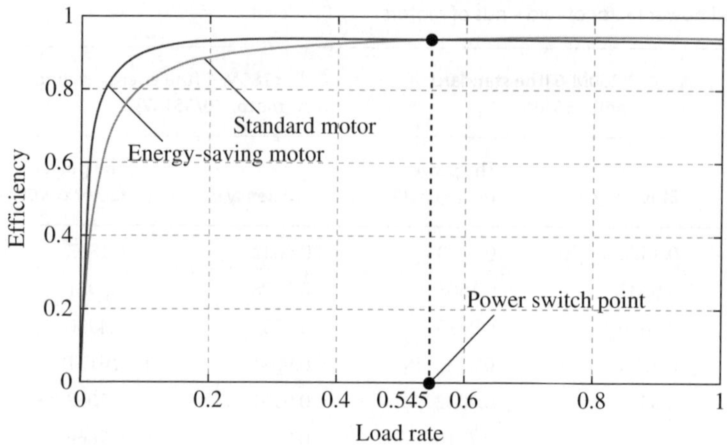


**Figure 4.3** The efficiency curve of two kinds of motors.

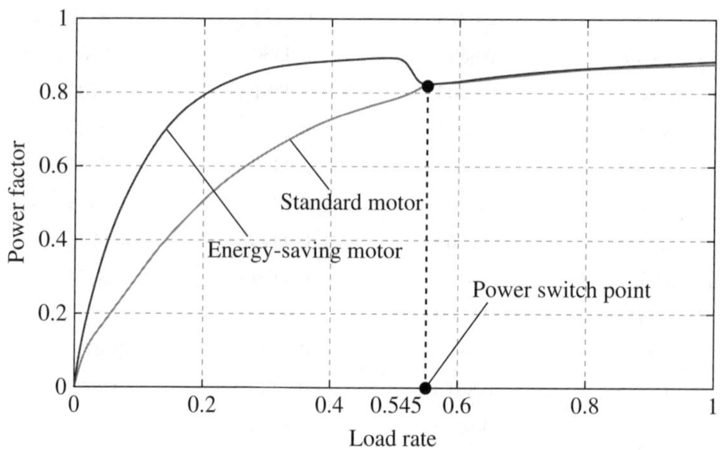


**Figure 4.4** The power factor curve of two kinds of motors.

factors of the area between the curve and the horizontal axis. In fact, using summation of the enough test points replaces the line integral along the curve.

#### 4.3.3.1 The Calculation of the Energy-saving Rate of the Average Active

Compare active power of the multiple operating point of the energy-saving motor with correspondences of the standard motor, then average their answers.

$$\Delta E_{\mathrm{p}} = \frac{1}{n}\Sigma\frac{P_i - P_i'}{P_i}\times 100\% \tag{4.8}$$

where $\Delta E_{\mathrm{p}}$, the rate of energy saving of the average active;
$P_i$, the input active power of the test point $i$ of the standard motor (kW);
$P_i'$, the input active power of the test point $i$ of the energy-saving motor (kW).

The efficiency of each test point into relationship, there is

$$\Delta E_{\mathrm{p}} = \frac{1}{n}\Sigma\frac{\eta_i' - \eta_i}{\eta'} \times 100\% \tag{4.9}$$

where $\eta_i'$, efficiency of the test point $i$ of the energy-saving motor;
$\eta_i$, efficiency of the test point $i$ of the standard motor.

Plug the date from Table 4.4 into the formula above, and obtain the rate of energy saving of the average active power.

$$\Delta E_{\mathrm{p}} = 7.03\%$$

#### 4.3.3.2 The Calculation of the Rate of Energy Saving of the Average Reactive

Compare the reactive power of multiple operating point of the energy-saving motor with correspondences of the standard motor, then gain the mean of their answers.

$$\Delta E_{\mathrm{Q}} = \frac{1}{n}\Sigma\frac{Q_i - Q_i'}{Q_i} \times 100\% \tag{4.10}$$

where $\Delta E_{\mathrm{p}}$, the rate of energy saving of the average reactive;
$Q_i$, the input reactive power of the test point $i$ of the standard motor (kW);
$Q_i'$, the input reactive power of the test point $i$ of the energy-saving motor (kW).

For each test point, the reactive power is that of active power; multiply $\tan\varphi$, then considering the relation between the active power and efficiency, there is

$$\Delta E_{\mathrm{Q}} = \frac{1}{n}\Sigma\frac{\eta'\tan\varphi_i - \eta\tan\varphi_i'}{\eta'\tan\varphi_i} \times 100\% \tag{4.11}$$

where $\tan\varphi_i'$, the tangent value of the phase angle of voltage and current of the test point $i$ of the energy-saving motor;
$\tan\varphi_i$, the tangent value of the phase angle of voltage and current of the test point $i$ of the standard motor.

The relationship of the trigonometric function $\tan\varphi = \sqrt{1-\cos^2\varphi}/\cos\varphi$.
Plug into the formula above, and you can see

$$\Delta E_{\mathrm{Q}} = \frac{1}{n}\Sigma\left(1 - \frac{\eta_i\cos\varphi_i\sqrt{1-\cos\varphi_i'^2}}{\eta_i'\cos\varphi_i'\sqrt{1-\cos\varphi_i^2}}\right) \times 100\% \tag{4.12}$$

where $\cos\varphi_i'$, the power factor of the test point $i$ of the energy-saving motor;
$\cos\varphi_i$, the power factor of the test point $i$ of the energy-saving motor.

Plug the date from Table 4.4 into the formula above, and get the rate of energy saving of the average reactive power.

$$\Delta E_{\mathrm{Q}} = 52.12\%$$

#### 4.3.3.3 The Calculation of the Average Comprehensive Rate of Energy Saving

From Formulas (4.9) and (4.12), obtain the average comprehensive the rate of energy saving

$$\Delta E_c = \Delta E_p + K_Q \times \Delta E_Q = 7.03\% + 0.09 \times 52.12\% = 11.72\%$$

In order to investigate the energy-saving potential of the double-power series winding motor, $K_Q$ is given 0.09.

The average comprehensive rate of energy saving of the dual-power series winding motor is 11.72%. It can be obviously seen that the energy-saving effect of the dual-power series winding motor is objective. If the original load is under 30 kW, the energy-saving effect of the dual-power series winding motor would be better.

### 4.3.4 The Control Method of the Dual-Power Series Winding Motor

In the practical application of the dual-power motor, the series winding switching must implement the automatic control. In order to realize automatic control, first, the voltage and current of the motor should be collected, computing the input power of the motor, and approximately calculating the load power of the motor. Then, load power determines the switching of dual-power winding. The test and calculation method are complex along with the difficulty of implementation and the high cost. Usually, according to the current working characteristic curve of the motor, as the curve of the $I_1 = f(P_2)$ judging the load power, automatic switching of the winding is realized. The principle of control system of the series dual-power winding is in Figure 4.5. Part ①

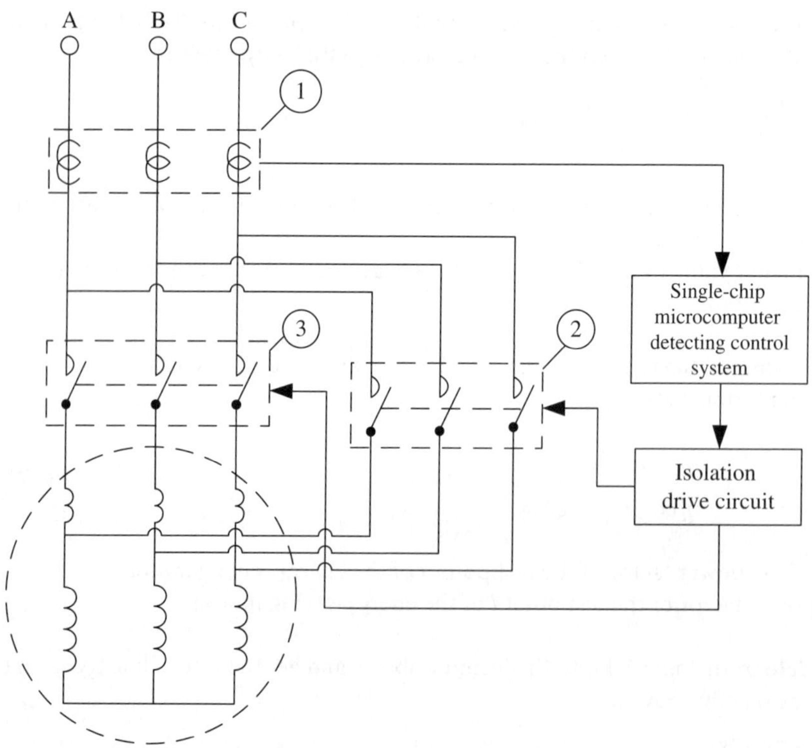


**Figure 4.5** The principle of control system of the series dual-power winding motor.

is the current transformer. Part ② is alternating current contactor of controlling the high-power winding, and Part ③ is alternating current contactor of controlling the low-power winding. The core of the motion control system is a single-chip microcontroller along with isolating driver circuit and so on. In the process of motor running, the microcontroller detects the working current of the motor through the current transformer. In the database to find the corresponding output power, judge the load of the motor and decide to put into the winding.

It can be assumed that the model of the motor is Y280M-6, its high-power winding being 55 kW and low-power winding, 30 kW. During motor starting, first, alternating current contactor ② closes and alternating current contactor ③ opens. If the detecting current shows that the output power of the motor current is less than 30 kW, the alternating current contactor ② disconnects and alternating current contactor ③ quickly closes. The two AC contactors cannot be closed at the same time; as since the small winding bears the overvoltage it will be easy to burn, so the two AC contactors in the circuit must interlock. In the actual circuit, there are the single chip microcomputer systems of power supply circuit and circuits for voltage detection and so on. The winding switching time should be controlled in 0.2 s. The switching cannot spend too much time, in order to avoid a larger decline in motor speed and the large shock current when AC contactor closes.

## 4.4 The Energy-saving Method of the Y-△ Conversion

Y-Δ conversion of windings is often used in reduced-voltage starting, where the starting current is reduced to a third of the original. In fact, Y-Δ conversion of windings is the applicability of the dual-power energy-saving motor. More than 3 kW motor winding commonly used Δ connection. If load is less than a third of the rated power in operation, the motor should be stopped, and open the junction box with Δ changing to Y connection. Or adopt the method of automatic switchover. When the load is lighter, the current greatly reduces after the motor winding switch into Y. Efficiency and the power factor are obviously improved, so as to achieve the goal of energy saving.

### 4.4.1 The Power Relations of Y-△

In the process of motor winding Y-Δ conversion, the motor current and the power significantly changes. Assume that the phase voltage of the motor winding in Δ connection is $U_\varphi$, and winding phase current is $I_\varphi$.

When the winding is Δ connection, the motor rated input power is

$$P_{1\Delta} = 3U_\varphi I_\varphi \cos\varphi_N \tag{4.13}$$

The motor rated output power is

$$P_{\Delta N} = P_{1\Delta} \times \eta_{\Delta N} \tag{4.14}$$

where $P_{\Delta N}$, the motor rated output power in Δ connection (kW);
$\eta_{\Delta N}$, the rated efficiency in Δ connection.

When the winding is in Δ connection, the winding phase voltage is $U_\varphi/\sqrt{3}$, the winding phase current is $I_\varphi/\sqrt{3}$.

In Δ connection of windings, the motor rated input power is

$$P_{1Y} = 3\times\frac{U_\varphi}{\sqrt{3}}\times\frac{I_\varphi}{\sqrt{3}}\cos\varphi_N = \frac{1}{3}\times 3U_\varphi I_\varphi\cos\varphi_N \tag{4.15}$$

After converting the winding to Y-Δ, the winding parameters is invariant; therefore, the power factor is same, so

$$P_{1Y} = \frac{1}{3}P_{1\Delta} \tag{4.16}$$

The motor rated output power in Y connection is

$$P_{YN} = P_{1Y}\times\eta_{YN} \tag{4.17}$$

where $P_{YN}$, the motor rated output power in Y connection;
$\eta_{YN}$, the rated efficiency in Y connection.

To compare (4.14) and (4.17), there is

$$P_{YN} = \frac{1}{3}\times\frac{\eta_{YN}}{\eta_{\Delta N}}P_{\Delta N} \tag{4.18}$$

In general, the efficiency of the low-power motor is lower than the efficiency of the high-power motor. Such as in Table 2.11, the efficiency of the Y series 4-pole 15 kW motor is 88.5%, and the efficiency of the Y series 4-pole 45 kW motor is 92.3%. The motor efficiency in Y connection is lower than the motor efficiency in Δ connection in series, because the mechanical loss is higher in the motor Y connection. So, in general, the power relation of motor winding Y-Δ conversion is as follows:

$$P_{YN} \approx 0.3P_{\Delta N} \tag{4.19}$$

From the above analysis, the original motor is the Δ winding, only when the load rate is less than 30%, under the condition of the "lighter load"; using the method of Y-Δ conversion increases the load rate to achieve the energy-saving goal. This method can be avoided if the load rate is 30–60%; otherwise, this makes the energy-saving method of Y-Δ conversion of motor winding subject to larger limit.

### 4.4.2 The Energy-saving Effect of Y-△ Conversion

#### 4.4.2.1 Loss Analysis

1) Decreasing Iron Loss. After the motor winding is converted from Δ to Y, each phase winding voltage of the stator drops down to about $1/\sqrt{3}$ of the rated voltage. Flux per pole is proportional to the voltage. There is

$$U_1 \approx E_1 = 4.44 f_1 N_1 k_{w1}\Phi_m \tag{4.20}$$

Every phase winding voltage drops down to about $1/\sqrt{3}$ of the rated voltage, and flux per pole drops to about $1/\sqrt{3}$ of the original flux per pole. Because the tooth of the motor slot size did not change, the magnetic flux density and iron loss decreases.

2) Decreasing Copper Losses of the Stator. After the motor winding is converted from Δ to Y, winding phase current drops down to $1/\sqrt{3}$, and the excitation current has dropped substantially. So the copper loss of stator decreases.
3) Copper Loss of Rotor Is Primitively Invariant. After the motor winding is converted from Δ to Y, the output power is invariant and the slip ratio as well as the electromagnetic power are primitively invariant. By the theory of motor to be seen,

$$p_{\mathrm{Cu2}} = sP_{\mathrm{em}} \tag{4.21}$$

So the copper loss of rotor is primitively invariant.
4) Mechanical Loss Is Primitively Invariant. After the motor winding is converted from Δ to Y, rotation rate is primitively invariant. So the mechanical loss is primitively unchangeable.
5) The Stray-Load Loss Is Changeless. According to the national standards, (GB/T1032-2005), the stray-load loss of the motor is proportional to output voltage:

$$P_s = P_1 \times \left[0.025 - 0.005\lg\left(P_2\right)\right] \tag{4.22}$$

After the motor winding is converted from Δ to Y, the output power is invariant, and the change of the input power is no more than 15%. So the stray-load loss changes are small.

Considering the above analysis of various losses, the principle of the energy-saving effect of Y-Δ conversion can reduce iron loss and stator copper loss.

#### 4.4.2.2 Testing and Analyzing Energy-saving Effect

On the test stand of the motor, test and analyze the Y-Δ conversion energy-saving effect for the three-phase induction motor (Y100L-4). This motor is 3 kW, the Δ connection, and the rated voltage of 380 V. The first to test for Δ connected in lighter load and smoothly adjusted the load into 0.7, 0.8, 0.9, 1, and 1.2 kW operation, respectively. Measure input active power, reactive power, and power factor, and calculate the efficiency. The testing and calculating results are listed in Table 4.5. After shutdown, open the junction box, and turn the Δ connection into the Y connection. Then test for the Y connection in the same way. The measured input active power, reactive power, the power factor, and calculated efficiency are recorded in Table 4.6.

As seen in Tables 4.5 and 4.6, when motor load is 0.7, 0.8, and 0.9 kW, which is the lighter load for the Δ connection motor (the rated is 3 kW), the efficiency is lower (ranging between 0.733 and 0.763), and the output power is higher. For winding in the Y connection (the rated power is 0.9 kW), the motor is close to fully load, the efficiency

**Table 4.5** Input active power, reactive power, the power factor, and efficiency of the Δ connection motor.

| | | | | | |
|---|---|---|---|---|---|
| $P_2$ (kW) | 0.7000 | 0.8000 | 0.9000 | 1.0000 | 1.2000 |
| $P_1$ (kW) | 0.9534 | 1.0571 | 1.1801 | 1.2859 | 1.4839 |
| $Q$ (kvar) | 2.0732 | 2.0658 | 2.1688 | 2.1788 | 2.0675 |
| Power factor | 0.3895 | 0.4290 | 0.4787 | 0.5090 | 0.5621 |
| Efficiency $\eta$ | 0.7330 | 0.7540 | 0.7630 | 0.7790 | 0.8090 |

**Table 4.6** Input active power, reactive power, the power factor, and efficiency of the Y connection motor.

| | | | | | |
|---|---|---|---|---|---|
| $P_2$ (kW) | 0.7000 | 0.8000 | 0.9000 | 1.0000 | 1.2000 |
| $P_1$ (kW) | 0.8448 | 0.9650 | 1.0799 | 1.2317 | 1.4875 |
| $Q$ (kvar) | 0.6529 | 0.6858 | 0.7191 | 0.7711 | 0.8740 |
| Power factor | 0.7867 | 0.8151 | 0.8329 | 0.8476 | 0.8622 |
| Efficiency $\eta$ | 0.8240 | 0.8290 | 0.8250 | 0.8140 | 0.8030 |

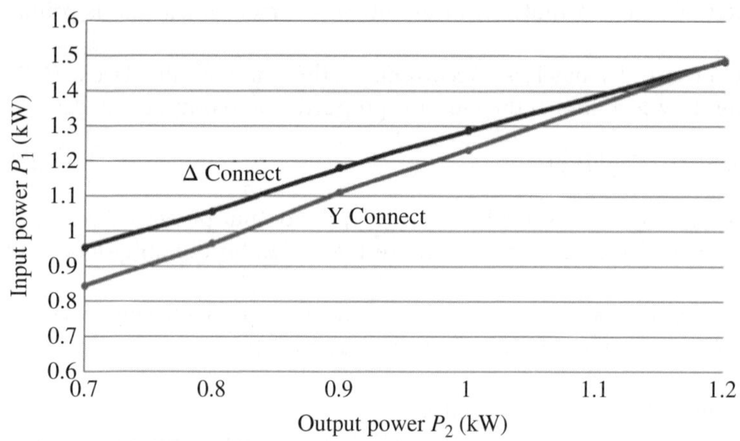


**Figure 4.6** The input power curve in the motor windings of Δ connection and Y connection.

is higher (ranging between 0.824 and 0.829), and the input power is lower. In 0.7–0.9 kW, when in Δ connection the average input power is higher than the Y connection by 0.1 kW. We can conclude that when the load is light, adopting the Y-Δ conversion method can achieve about 10% of the energy-saving effect. Meanwhile, adopting this method can make the reactive power drop down to about 60% and the power factor increases about 45%.

When the load is added up to 1 kW, the energy-saving effect drops down to about the 5%. When the load is added up to 1.2 kW, the output power in Y connection is slightly higher than it in Δ connection. In both cases the input power along with the load change is shown in Figure 4.6. It is not difficult to find that the Y connection should not exceed 30% of the motor rated power. Otherwise, the energy-saving effect gradually disappears, and the motor easily burns.

### 4.4.3 The Y-Δ Conversion Control Circuit

As with control circuit of the dual-power motor above, the winding Y-Δ conversion also needs automatic control. First collect the motor current, and then according to the current of the motor characteristic curve, determine the output power of the motor, judge the load rate, and realize the winding Y-Δ automatic switching. The principle of control system of the motor winding Y-Δ conversion is shown in

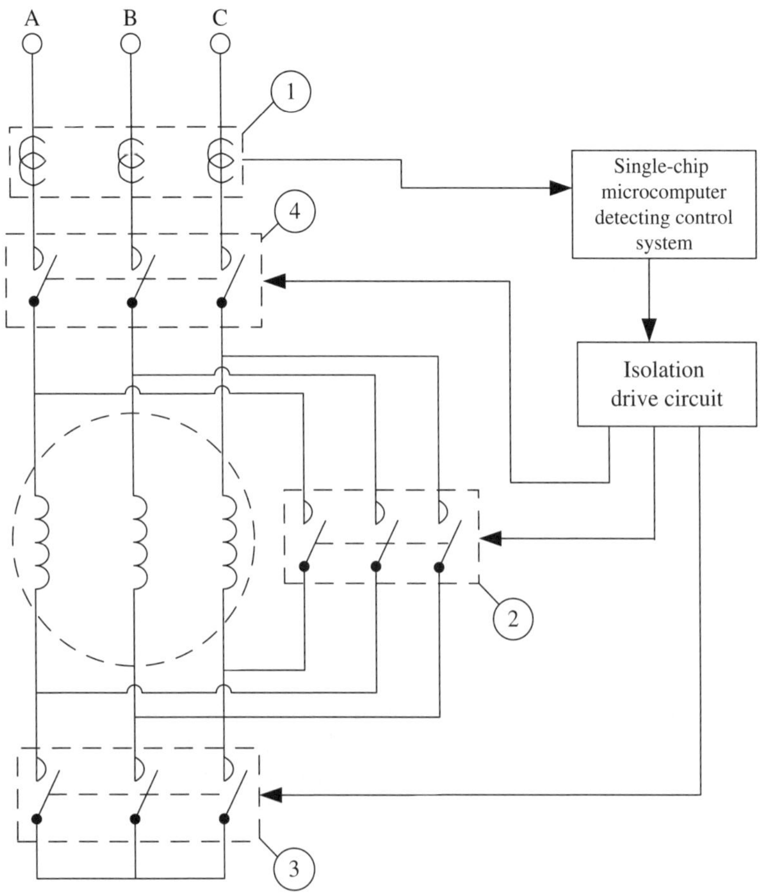


**Figure 4.7** The principle of control system of the motor winding in Y-Δ connection.

Figure 4.7. Part ① is the current transformer, part ② is the alternating current contactor of controlling winding in Δ connection, part ③ is the alternating current contactor of controlling winding in Y connection, and part ④ is the alternating current contactor of controlling the power on and off in this figure. In addition, there is the single chip microcontroller, driver circuit of contactor, and so on. In the process of the motor running, the single chip microcontroller detects the working current of the motor through the current transformer. In the database to find corresponding output power of the motor, judge the load of the motor and decide whether it compliments the Y-Δ connection. As you can be seen from Figure 4.7, when the alternating current contactor ② separately closes, the motor winding switches the Δ connection. The alternating current contactor ④ is power master switch. The two AC contactors cannot be kept closed at the same time; otherwise it will cause a short circuit of power supply, so the two ac contactors in the circuit must interlock. It is necessary to set up a power master switch, in order to avoid motor charged after shutdown. In the actual circuit, there are power supply circuits for the single chip microcomputer system, circuits for voltage detection, and so on.

The things requiring attention are that in the process of motor winding Y-Δ connection, the phase sequence ensures the same. In control system above, the head phase sequence of three-phase winding in Y connection is A → B → C. The end of A phase connects the head of B phase, the end of B phase connects the head of C, and the end of C phase connects the head of A phase. So the head phase sequence of three-phase winding in Y connection is still A → B → C.

## 4.5 The Energy-saving Method of Extended △ Winding Switching

### 4.5.1 The Design Principle of the Extended △ Winding

The motor Y-Δ conversion is extremely limited, if the load rate is less than 30%, Y-Δ conversion cannot be used. In order to solve the problem, a common method of dual-power wingding motor design is the extended Δ winding. The connection principle of the extended Δ winding is shown in Figure 4.8. Every phase winding has a tap. If taps are first connected, the whole winding is Δ connection, as shown in Figure 4.8a. This connection is the same as Δ connection of the Y-Δ connection, and the motor outputs high power. If the end of taps and winding are connected into Δ, the other part forms the extended side of the Δ, as shown in Figure 4.8b, that is the method of extended Δ winding. The extended Δ winding is low power.

The extended Δ winding power allocation is related to voltage. In Figure 4.8b, the Δ winding share more voltage, as well as high power. In the same winding, the Δ in the extended Δ is bigger, and the extended Δ winding power is higher; otherwise the extended Δ winding power is lower. In general, the two powers of a dual-power motor (the extended Δ winding and series winding of dual-power) should conform to the power levels of the state standards. Such as, the high power is 30 kW, so the low power can be chosen 22 or 18.5 kW; the high power is 45 kW, so the low power can be chosen 37 kW or 30 kW. The high power is generally the original motor power, and the low power is depended on the variation range of the load.

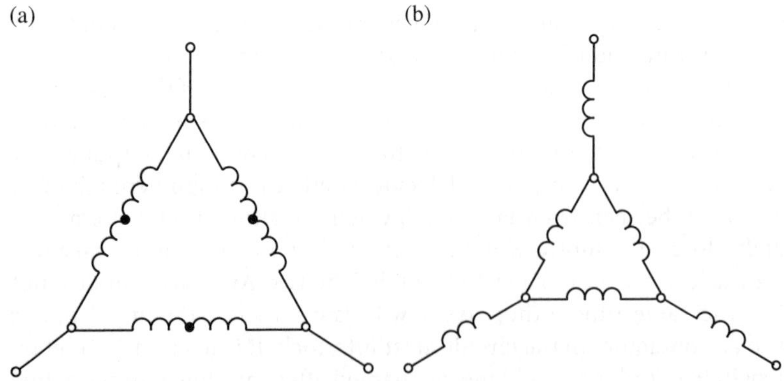


Figure 4.8 The connection principle figure of the extended Δ. (a) Δ connection and (b) extended Δ connection.

### 4.5.2 The Switching Control Circuit for the Extended △

The principle of switching control circuit system for the extended Δ winding is shown in Figure 4.9. In this figure, part ① is the alternating current contactor, part ② is the alternating current contactor of controlling the Δ winding, part ③ is the alternating current contactor of controlling the extended Δ winding, and part ④ is the alternating current contactor of controlling the power on and off. In addition, there is the single chip microcontroller for detecting, isolating driver circuit of contactor, and so on. From the figure, when AC contactor ② is closed, the whole winding is connected into big Δ; when AC contactor ③ is closed, the below winding is connected into small Δ, and the below winding forms the extended side of small Δ. The two AC contactors cannot be closed at the same time; otherwise it will burn the winding of the extended side part, so the two AC contactors in the actual circuit must interlock. The AC contactor ④ is the master power switch. If there is no master power switch, the motor will

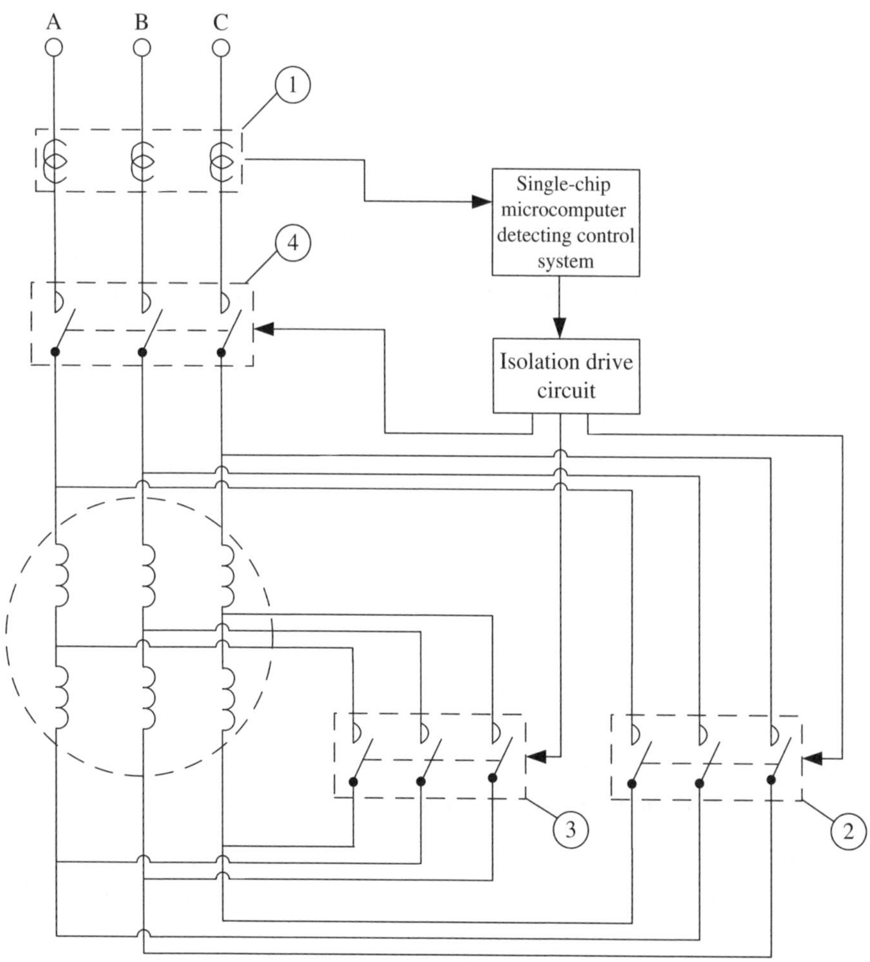


**Figure 4.9** The principle of control system of the extended Δ winding switching.

charge after shutdown, which is unsafe. In the process of the motor running, the microcontroller detects the working current of the motor through the current transformer. In the database, to find the corresponding output power of the motor, judging the load rate decides whether the winding will switch or not. In the actual circuit, there are power supply circuits for a single chip microcomputer system, circuits for voltage detection, and so on.

### 4.5.3 The Comparison of Dual-Power Motor

The dual-power motor is the main method for the motor energy-saving reconstruction. In the three dual-power motors introduced above, the following analyzes and compares their advantages and disadvantages.

#### 4.5.3.1 Power Range

From coverage of power aspects, series winding and extended Δ winding dual-power motors have wide applicability, but the Y-Δ conversion has a narrow applicability. The series winding and extended Δ winding dual-power motor is more applied.

#### 4.5.3.2 Winding Design and Manufacturing Cost

From winding design and manufacturing aspects, the Y-Δ conversion dual-power motor is the most simple. As long as the original winding is Δ, and the voltage is 380 V, it is convenient to realize double power, and no change in the winding required. Series winding is slightly simpler than extended Δ winding, but the extended Δ winding has high material utilization. In a high–power motor, its efficiency is higher than series winding for dual-power motor. In a low power motor, these two kinds of motor efficiency are close. The improvement cost is the same.

#### 4.5.3.3 The Cost of Control System

From Figures 4.5, 4.7, and 4.9, you can see that the control system of series winding dual-power just need two AC contactors, but the control system of the other two kinds of dual-power motors need three AC contactors. Relatively speaking, for the control system of series winding dual-power motors, its cost is lower, and its reliability is slightly higher.

In terms of the design dual-power motor, there are also other methods, such as double Y type, the Δ embedded Y type, and so on. But, at point of power coverage, winding design, processing cost, and the cost of control system, other methods lose advantages. So it is very important to master the principle and control method of series winding, extended Δ winding and Y-Δ conversion.

## 4.6 Discussion Topics in the Chapter

1) What is the delimiter of three-phase induction motor “lighter load?”
2) Why does three-phase induction motor consume electricity in lighter load?
3) What is the principle of three-phase induction motor power matching?
4) What are the basic steps of three-phase induction motor?
5) What are dual-power motors, and what are its advantages?

6) In general, what kinds of methods do the motor have in the winding of dual-power?
7) The Y-Δ conversion of motor winding is also a dual-power motor, and what are its advantages and disadvantages?
8) In general, how is power relationship of the motor winding in Y-Δ conversion?
9) How are the connection forms of the extended Δ winding?
10) What are the advantages and disadvantages of extended Δ winding dual-power motor?

# 5

# Energy-saving Principle and Methods of Speed Matching

In order to meet the requirements of production process in production practice (mine ventilation system, residents' water supply system, central air conditioner, etc.), the motor speed should be regulated. The electricity consumption of this kind of production machinery is very large. According to the statistics, the electricity consumption of pumps accounts for 21% and fans account for 12% of the total power consumption in China. Due to the constraints in design, there is no motor speed regulating system in some fans and pumps. When the demands for wind or water change, the constant-speed motor cannot meet the requirements, and the motor is in a state of production capacity surplus, wasting a lot of electricity. So the researches for the energy saving of fan and pump systems are very important. In order to study the energy saving of fan and pump systems, this chapter will first present the energy-saving principle of speed matching and take the water supply system as case to introduce the energy-saving theory analysis of pump speed regulation and control principle of constant pressure water supply system. Then, the basic principle of variable frequency speed control, the selection, and application examples of frequency converters is presented. Finally, it will introduce the pole changing adjustable speed method of an asynchronous motor. The energy-saving principle and application of combined pole changing adjustable speed motor in the multiple pump system is analyzed.

## 5.1 Energy-saving Principle of Speed Matching

The power system comprising a fan and a pump can be referred to as the fluid delivery system, which is one of the main research fields of motor energy saving. In the fluid delivery system, the fluid mechanics equations of fan and pump are basically the same, so the two systems can be bracketed together. For the study of the energy-saving principle of pump speed matching, the basic parameters of the pump are described below.

### 5.1.1 Basic Parameters of the Pump

Water pump is a kind of mechanical equipment to convey fluid, and the pump consists of a vane pump and a positive displacement pump. Relying on the dynamic effect of a rotating impeller on liquid, a vane pump conveys energy to liquid continuously, increasing the kinetic energy and pressure energy of liquid, and then convert kinetic energy

*Energy-saving Principles and Technologies for Induction Motors*, First Edition. Wenzhong Ma and Lianping Bai.

into pressure energy in the delivery chamber. The vane pump includes centrifugal type, axis-flow type, mixed-flow type, and so forth. A positive displacement pump conveys energy to liquid periodically through periodic changes of working volume, increasing the liquid pressure, and then the liquid is drained away forcibly. The positive displacement pump includes a piston pump, diaphragm pump, and so on. Since the piston pump is applied more extensively in pumps, the following part will introduce the basic parameters of the piston pump.

1) *Quantity of flow*
   Quantity of flow is the quantity of pumped liquid in unit time, commonly known as volume flow, expressed by $Q$, and its unit is $m^3/s$.
2) *Head*
   Head is the liquid column height of the pumped liquid, and its unit is m. The water level variation (namely the corresponding water level difference) from one location to another location is customarily expressed as head.
3) *Total Head $H_T$*
   Total head, also called full head, is a physical quantity representing the pumping ability, including the required head of lifting water from the pool to the highest level, overcoming pipe resistance, and keeping flow velocity. Total head numerically equals the maximum height of the pump, regardless of pipe resistance and flow velocity.
4) *Actual Head $H_A$*
   Actual head refers to the actual required head of lifting water level by pump, and it shows the actual height of lifting water level by pump, so it is called actual head.
5) *Loss Head $H_L$*
   Loss head is the required head to overcome pipe friction loss and other losses during supplying water and the required head to make water flow with a certain velocity. The relation of $H_T$, $H_A$, and $H_L$ is $H_T = H_A + H_L$.
6) *Torque of Centrifugal Pump*

$$T_L = T_0 + K_T n^2 \tag{5.1}$$

   where $T_L$, the pump torque (Nm);
   $T_0$, the no-load torque (Nm);
   $n$, the pump speed (r/min);
   $K_T$, the torque coefficient.

Mechanical characteristic curve of the pump is shown in Figure 5.1.

In Equation (5.1), no-load torque $T_0$ is the loss torque of the pump itself, mainly formed by friction and other resistance. Compared to the rated torque, the no-load torque is very small and may be neglected in most cases, so Equation (5.1) can be simplified as

$$T_L \approx K_T n^2 \tag{5.2}$$

Centrifugal pump is often referred to as square torque load, and this rule is the intrinsic characteristics of centrifugal pump; it does not vary with control method. In other words, no matter which control method is adopted in the water piping system, once the speed changes, blockage torque formed by counterforce of water flow to blade always conforms to this law.

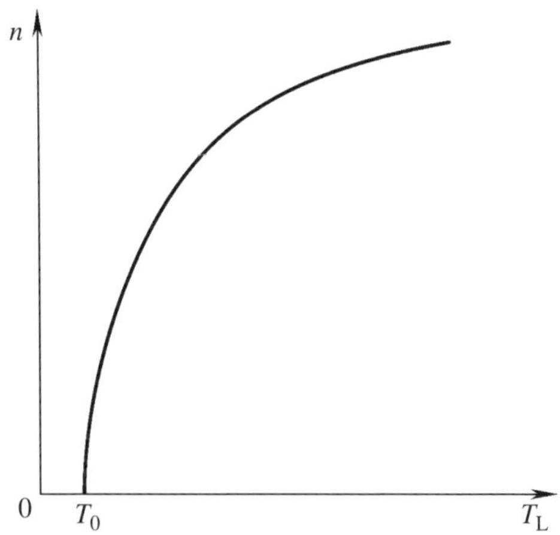


**Figure 5.1** Mechanical characteristic curve of pump.

### 5.1.2 Energy Analysis of Water Supply System

Surplus head or surplus flow is the main reason of energy wastage in the fluid transmission system (i.e., supply exceeding demand), and the second reason is that the equipment operation efficiency in fluid transmission system is very low.

For the fluid transmission system, speed control scheme is to adjust the relationship of supply and demand. So if the supply and demand of the original constant speed fluid transmission system are equal, it is meaningless to add a speed control device. Moreover, all speed control devices have energy losses. From qualitative analysis, it can be known that the amount of energy saving of applying speed control device is equal to the energy consumption reduction of the surplus head (or surplus flow) minus the energy consumption of the speed control device.

To illustrate this problem, an example will be given as follows: A student dormitory building water supply system is shown in Figure 5.2. In the figure, 1 is the motor, 2 is the pump, and 3 is the check valve. Each layer of the dormitory building has a laundry room. The height of the dormitory building, $H$ is 31 m. The peak flow of morning, noon, and afternoon should be met in design. The peak value of flow demand is 50 m$^3$/h and the valley value is 40 m$^3$/h. The pump is a centrifugal pump, the rated speed is 2900 r/min, the flow at rated speed is 50.3 m$^3$/h, and the shaft power of the pump at rated speed is 5.85 kW. The pump flow satisfies the requirement of demand at the peak. Since the pump is relatively far away from the dormitory building, and takes the network characteristics of water duct into consideration, the effective head of pump is determined to be 60 m. The rated efficiency is 79.17%.

The rated power of the motor is 7.5 kW, the rated speed is 2900 r/min, and the rated efficiency is 87.0%.

The speed-regulating problem has not been taken into consideration in design. Without a speed control device, the speed of motor keeps constant and the pump flow is unchanged; when the actual water usage decreases, the head increases, and the pressure in the pipe increases too. At that moment, a humming chatter of pipes will be heard when opening the faucet. Thus, the input power of the motor is almost unchanged, which would cause a waste of electricity. If the motor uses a speed control system, at the valley usage, the speed of the motor and the pump flow decrease, under the condition

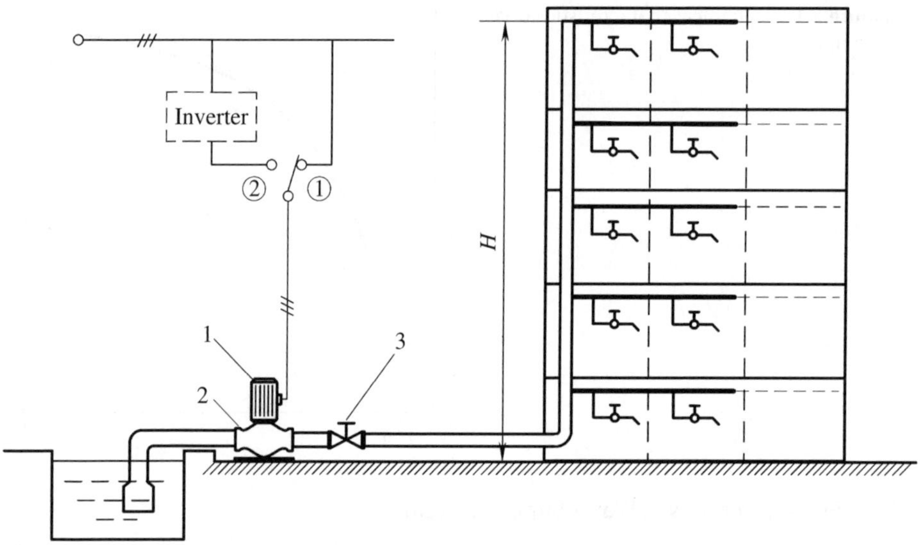


Figure 5.2 A diagram of building water supply system. 1, Motor; 2, pump; 3, check valve.

of assurance of unchangeable water supply pressure, the head keeps unchanged, and the input power of the motor will decrease.

Below is an example of common centrifugal pump and the relationship of the shaft power and speed of the pump will be discussed.

Pump flow is proportional to the speed

$$Q = Q_N \frac{n}{n_N} \tag{5.3}$$

where $Q$, the pump flow ($m^3/h$);
$n$, the pump speed (r/min);
$Q_N$, the pump flow at rated speed ($m^3/h$);
$n_N$, the rated speed of pump. $Q_N$ and $n_N$ are constants.

Pump head is proportional to the square of the speed:

$$H = H_N \left( \frac{n}{n_N} \right)^2 \tag{5.4}$$

where $H$, the pump head (m);
$H_N$, the pump head at rated speed (m), which is a constant.

Shaft power of the pump

$$P_{sb} = k_p \frac{QH}{6.12\eta_p} \tag{5.5}$$

where $P_{sb}$, the shaft power of the pump (kW);
$\eta_p$, the rated efficiency of the pump;
$k_p$, a constant.

Substituting (5.3) and (5.4) into (5.5), the shaft power can be derived as follows:

$$P_{sb} = k\left(\frac{n}{n_N}\right)^3 \tag{5.6}$$

where $k$ and $n_N$ are both constants.

The rated pump efficiency is associated with the change of the speed. When the speed has only a little change near the rated speed, the change of pump efficiency along with the change of speed could be ignored and the shaft power of the pump is proportional to the square of the speed.

So, when the flow demand decreases, the speed drops in proportion, while the shaft power decreases with cubic proportion. Without a speed control device, both the shaft power of the pump and output power of the motor remains unchanged, so the input power of the motor is constant. This is the main reason of energy wastage which is caused by the constant speed operation of the motor. In this case, the motor and the pump will produce excess energy, increasing the pressure of the pipe. The common solution is to install a throttle valve or a turn-on back-flow, which is actually a waste of the surplus flow.

In the fluid transmission system, when the demand is less than the supply, and if the motor operates at a constant speed, such as a three-phase asynchronous motor running under power frequency, it will produce excess energy for the motor, rather than lower load-carrying. The solution for the problem of lower load-carrying does not apply to this condition. In addition, the methods of motor speed regulation should be adopted. Figure 5.2 will be taken for an example to analyze the energy loss caused by surplus flow at valley usage of water and the energy-saving rate of using variable frequency speed controlling.

#### 5.1.2.1 Energy Consumption of Motor in Constant Speed Operation

In Figure 5.2, there is no speed control device in the motor, the switch is at ① position. In this case, the motor and the pump are basically running at the rated point. The input power of the pump is 5.85 kW, and it is also the output power of the motor. The rated efficiency of the motor is 0.87, and the input power of the motor is 6.724 kW.

#### 5.1.2.2 Energy Consumption of Motor in the Variable Frequency Speed Control Operation

In Figure 5.2, the motor installing a variable frequency speed control device, the switch is at ② position. At valley usage of water, when the water usage flow reduces to 40 $m^3$/h, the flow falls down about 20%. Adjusting the output power of the frequency converter can make the motor speed fall down 20%, since the shaft power is proportional to the cube of the speed, namely, the shaft power decreases 48.8%. So the efficiency of the pump keeps unchanged. The shaft power of the pump is 2.995 kW, and the output power of the motor is 2.995 kW. The efficiency of the motor keeps unchanged, so the input power of the motor is 3.443 kW. The efficiency of the selected frequency converter is 90%, and the input power is 3.825 kW. The energy consumption of the frequency converter itself is 0.383 kW.

The surplus flow at valley usage of water is 20%, and the shaft power is proportional to the cube of the speed, so the power $\Delta P'$ consumed by the surplus flow equals to 48.8% of the input power of the motor in constant speed, that is,

$$\Delta P' = P_1 \times (1 - 0.8^3) = 6.724 \times 0.488 = 3.281\text{kW}$$

In fact, the saving power of the motor by using a speed control device equals to the power consumed by surplus flow minus the power loss of the frequency converter itself, that is,

$$\Delta P = \Delta P' - \Delta P_{\text{BF}} = 3.281 - 0.383 = 2.898\text{kW}$$

The saving power of the motor by using a speed control device can also be obtained by subtracting the input power of the frequency converter in speed control from the input power in constant speed, that is,

$$\Delta P = P_1 - P_1' = 6.724 - 3.825 = 2.899\text{kW}$$

#### 5.1.2.3 Power-Saving Rate of Using Variable Frequency Speed Control

The power-saving rate of variable frequency speed control of motor is

$$\Delta E = \frac{\Delta P}{P_1} = \frac{2.899}{6.724} \times 100\% = 43.1\%$$

It can be seen that when the demand of water usage declines, the energy-saving potential using a variable frequency speed control to decrease speed is very large. But it does not mean that the decrease of the shaft power is the amount of the saving energy; the loss of the frequency converter itself should also be considered. Strictly speaking, the decrease of the pump efficiency caused by the decrease of the speed should also be taken into account.

In the water supply system, the power of the motor, the power of the pump, and the rated speed are determined according to the maximal flow and head. During the running process of the system, when the water usage decreases, the energy consumption caused by the surplus flow can be eliminated by adjusting the speed of the motor, and that is the energy-saving principle of motor speed matching. The motor energy saving of variable frequency speed control (or other speed control) and the constant-speed operation are relative. The quantity of saving potential is approximately equal to the surplus flow produced in constant-speed operation, and the energy loss of the control device should also be considered. There would be no saving potential if no surplus flow exists in the constant operation. Still equipped with a frequency converter, it would only consume more energy.

In addition to the need of energy saving, in the transmission system, using variable frequency speed control or other methods of speed control to adjust speed in real time and to track the change of the terminal demand is also the basic requirement of the production process.

### 5.1.3 Efficiency Analysis of Speed Control Water Supply System

From the building water supply system in Figure 5.2, it can be seen that the input energy of the power supply is mainly distributed into four parts, including frequency converter, motor, pump, and pipe network, as shown in Figure 5.3. In the figure, $P_1$ is the input power of the frequency converter, $P_2$ is the output power of the frequency converter and is also the input power of the motor; $P_3$ is the output power of the motor and is also the input power of the pump; $P_4$ is the output power of the pump and is also the input power of the pipe network; and $P_5$ is the output power of the pipe network passed on to the fluid, that is, converted to potential energy and kinetic energy of the fluid.

The input electric power of the frequency converter in every part of the system has loss. For further analysis of the energy-saving problem of the water supply system, the efficiency of every part of the water supply system should be discussed in detail. The energy distribution of the water supply system is shown in Figure 5.4. In this figure, $p_1$ is the loss of the frequency converter, $p_2$ is the loss of the motor, $p_3$ is the loss of the pump, and $p_4$ is the loss of the pipe network.

The efficiency of the frequency converter in Figure 5.4 is

$$\eta_1 = \frac{P_2}{P_1} \times 100\% \tag{5.7}$$

The efficiency of the frequency converter is related to the output frequency and load ratio, that is, the lower the output frequency is, the lower the efficiency will be, and the lower the load ratio is, the lower the efficiency will be.

Efficiency of the motor is

$$\eta_2 = \frac{P_3}{P_2} \times 100\% \tag{5.8}$$

Strictly speaking, the efficiency of the motor has a relationship with the speed, that is, when the speed (output frequency of the frequency converter) is low, the efficiency will

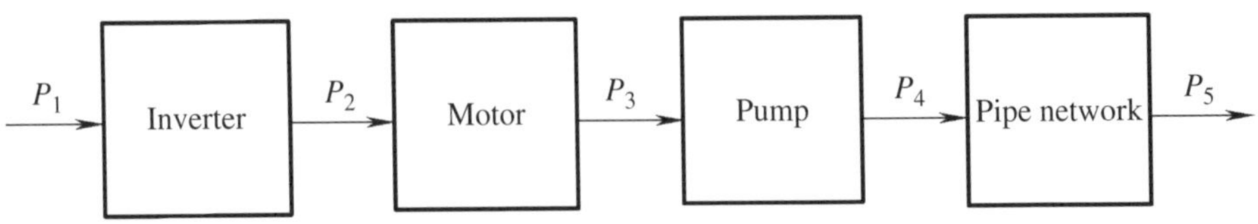


**Figure 5.3** Structure diagram of speed control water supply system.

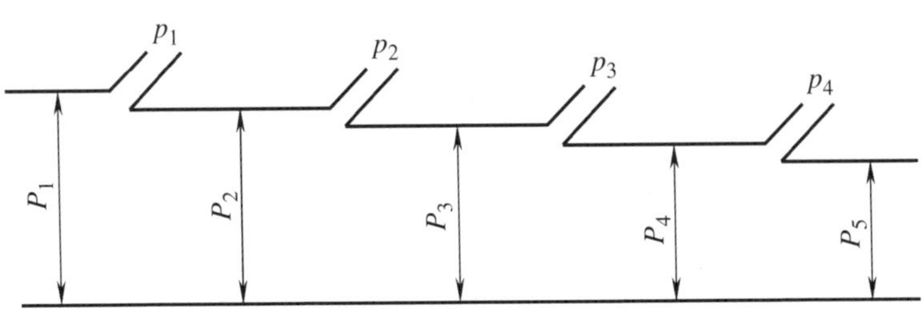


**Figure 5.4** Energy distribution of the water supply system.

decrease. The harmonic produced by the frequency converter would lead to the increase of the supplementary loss of the motor.

Pump efficiency is

$$\eta_3 = \frac{P_4}{P_3} \times 100\% \tag{5.9}$$

Pump efficiency is closely related to the speed, there is little change of the efficiency near the rated speed. If the speed falls down to about 60% of the rated speed, the efficiency would decrease a lot.

The efficiency of the pipe network is

$$\eta_4 = \frac{P_5}{P_4} \times 100\% \tag{5.10}$$

The efficiency of the pipe network is related to the network structure, pipe network resistance, and water velocity. The lager the pipe network resistance is, the lower the efficiency will be. When water velocity decreases, the efficiency would increase a little.

The overall efficiency of the speed control water supply system is

$$\eta = \eta_1 \times \eta_2 \times \eta_3 \times \eta_4 \tag{5.11}$$

In conclusion, energy saving of the water supply system is not only a problem of speed matching, the influence of the change of the speed on every part efficiency should also be taken into account. Therefore, the energy saving of the water supply system also includes the overall optimization design. Energy saving cannot be simply achieved by adding a frequency converter in front of the motor. For example, in Figure 5.2, if the public service facilities stop working, the flow demand would be less than 10 $m^3$/h at valley usage of water. If decreasing the pump speed by the frequency converter, the pump efficiency would be very low, harmonic loss increases a lot, the decline of the overall efficiency of the water supply system would be very large, and thus there would be a big discount in the energy-saving effect of the frequency converter.

### 5.1.4 Comparison of Various Motor Speed Control Methods

A motor speed control device is superior to the mechanical speed control device in speed control performance, efficiency, cost, and reliability, so the motor speed control device is extensively used in the speed control system. It is well-known that the main control methods of a three-phase asynchronous motor include variable frequency speed control, pole changing speed control, cascade speed control, and variable voltage speed control. In addition to variable frequency speed control, those speed control methods are developed with the development of power electronics, and each has its own merits. This section makes a brief analysis and comparison of them.

#### 5.1.4.1 Variable Frequency Speed Control

Variable frequency speed control is to change the power frequency, thus changing the synchronous speed of the motor. In order to make full use of the capacity of the motor in the process of speed control, the magnetic flux should be constant. When frequency decreases, the voltage must decrease with frequency, and the matching change of the voltage and the frequency is called coordination control of constant flux variable frequency speed control. According to different coordination control methods of voltage and frequency, the speed control methods of constant torque or constant power characteristic can be obtained. The features of variable frequency speed control are as follows:

1) The efficiency of speed control is high, and the supplementary loss of the motor is low in the process of speed regulation.
2) The range of speed regulation is very large, the speed regulation ratio can reach to 20:1 with hard mechanical properties and high precision.
3) The field of application is very large, widely used in a cage asynchronous motor.
4) The technology is complex, cost is high, and maintenance is difficult.

Variable frequency speed control is suitable for the occasions of high accuracy and good speed control performance, such as rolling roller which is synchronous driven by several motors, spinning machinery, fan, and pump.

#### 5.1.4.2 Pole Changing Speed Control

This method of speed control is changing the connection modes of winding to change the stator poles of the squirrel cage motor to regulate the speed. The characteristics of pole changing speed control are as follows:

1) High speed regulation efficiency, hard mechanical properties, and good stability;
2) Simple control equipment, convenience, and low cost;
3) Step speed regulation, big stairstep, smooth speed regulation is impossible.

Pole changing speed control is suitable for the production machinery or occasions that do not need stepless speed regulation, such as metal cutting, puddle mixer, lifter, hoisting equipment, fan and pump, and so forth.

#### 5.1.4.3 Cascade Speed Control

Cascade speed control is to add a device in series with the rotor loop of wound motor, which can adjust electric additional potential to change the slip of the motor and achieve the aim of speed control. Most of the slip power is absorbed by a series additional electric potential, and the absorbed slip power is fed back to the grid by the device which can produce an additional electric potential. The features of the cascade speed control are as follows:

1) The slip loss is fed back to grid in the process of speed control, leading to the higher speed control efficiency.
2) Device capacity is proportional to the speed range, applied for the situation that speed control range is 70–90% of the rated speed.
3) When speed regulating device breaks down, it can be switched to full speed mode, avoiding production stopping.
4) The power factor of thyristor cascade speed controlling system is low and the influence of harmonics is large.

Cascade speed control is suitable for fan, pump, rolling machine, mine hoist, and extruder.

**Table 5.1** Comparison of performance, efficiency, cost, and reliability of all kinds of speed control mode.

| Speed control mode | Device efficiency | Speed control range | Smoothness | Cost | Reliability | Maintainability |
|---|---|---|---|---|---|---|
| Variable frequency speed control | 0.9–0.95 | 20 : 1 | Good | Higher | Better | Better |
| Pole changing speed control | ≈1 | Small | Bad | Lower | High | Good |
| Cascade speed control | ≈0.9 | 2 : 1 | Better | Higher | Better | Better |
| Variable voltage speed control | ≈0.85 | 1.5 : 1 | Better | Higher | Better | Better |

#### 5.1.4.4 Variable Voltage Speed Control

Changing the stator voltage of the motor can get a set of different mechanical characteristic curves, as shown in Figure 2.17, thus a certain range of speed control can be obtained. Since the torque of the motor is proportional to the square of the voltage, maximum torque decreases a lot and the speed range is smaller. In fact, the variable voltage speed control device is a voltage-controlled power source. At present, common speed control methods include a series transductor, an autotransformer, SCR voltage regulation, and so on. The features of the variable voltage speed control are as follows:

1) Simple variable voltage speed control circuit, which is easy to realize automatic control, convenient maintenance;
2) Larger differential power in regulating process, which leads to low efficiency of the variable voltage speed control device;
3) Suitable for wound rotor induction motor, or specially designed YH series high resistance rotor motor, but be careful when used in ordinary squirrel cage motor.

Variable voltage speed control is suitable for the production machinery of small speed control scale below 100 kW.

The comparison of performance, efficiency, cost, and reliability of the above four kinds of speed control modes are shown in Table 5.1. It is not hard to see from the table that the highest comprehensive performance index is the variable frequency speed control. So the variable frequency speed control is the mainstream of the speed control of a three-phase asynchronous motor.

## 5.2 Energy-saving Theoretical Analysis of Pump Speed Control

### 5.2.1 Characteristic Curve of Pipe Network

Pipe network resistance is the physical quantity that indicates the resistance effect of a pipe network system to water flows. Its value in static state mainly depends on the structure of the pipe network, including the length and diameter of the pipe, resistance coefficient of each part of the pipe, and so on, and it is associated with the flow velocity, namely associated with the flow in dynamic state. Pipe resistance is difficult to calculate quantitatively by formula, usually described by the relationship curve of head and flow.

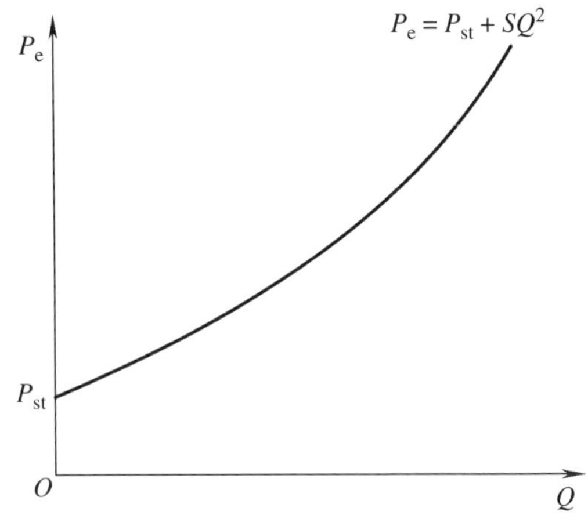


**Figure 5.5** Pipe network characteristic curve.

In the water supply system, the relationship of the required energy of flowing liquid $P_e$ in pipe network and flow $Q$ is

$$P_e + P_{st} + SQ^2 \tag{5.12}$$

where $P_{st}$ is the energy dissipation that the fluid overcomes in the static head of pipe network, which reflects the influence of the external environment on pipe network flow, including gravity and exchange surface pressure of the fluid in the pipe and external environment. When pipe network is in stable operation condition, $P_{st}$ has nothing to do with the change of the flow.

$S$ is the total resistance of pipe network. The relationship is described as a curve using flow as the abscissa, and pressure as the ordinate, namely, the pipe network characteristic curve (as shown in Figure 5.5).

The determinant that influences the shape of the pipe network characteristic curve is the total impedance of the pipe network $S$. The larger the value of $S$ is, the steeper the curve will be. When the pipe network system is installed, unchanging the valve opening, the length, diameter, and local resistance coefficient are constant, that is, $S$ is a fixed value. For a specific pipe network, its network characteristics are determined, usually using head to represent the energy that fluid flows through the pipe, namely

$$H = H_{st} + RQ^2 \tag{5.13}$$

where $H$, the fluid head (m);
$H_{st}$, the static head of the pipe network (m);
$R$, the coefficient of pipe resistance.

### 5.2.2 Pump Characteristic Curve

Pump characteristic curve describes the relationship of the main parameters, namely, head and flow, power and flow, efficiency and flow. Mathematical models of the relationships are very complicated; usually the test curves are given by manufacturers, or the main sections of the characteristic curve are described by simple mathematical formulas.

#### 5.2.2.1 Head–Flow Curve of Pump

There is a certain change rule between the head of pump and flow, which is related to the factors such as the structure of pump itself and so on. In normal use, the rule for a specified pump is fixed; only with the increase of service years, due to wear and corrosion of the pump impeller, flow channel of water pump housing, bearings and sealing element, this rule will be changed. This changing rule is called head–flow curve of pump, generally called the *H-Q* characteristic curve. The mathematical model that reflects this curve is rather complicated, and it is an irregular decline curve, which is difficult to be used in engineering design. It is usually described by approximate parabola, namely

$$H = AQ^2 + BQ + C \tag{5.14}$$

where $H$ is the pump head;
$Q$ is the pump flow;
$A$, $B$, and $C$ are the three coefficients of the parabola; once $A$, $B$, and $C$ are determined, the pump characteristic curve is determined.

Common head–flow curve of pump is shown in Figure 5.6 (Curve 1). Three working point $m_1$, $m_2$, and $m_3$ on the curve are usually given by manufacturers; these three points are the high-efficiency areas of the pump. Substituting the values of head and flow of these three points into (5.14), the three coefficients $A$, $B$, and $C$ can be solved, so the equation of pump head–flow characteristic curve can be obtained.

#### 5.2.2.2 Power–Flow Curve of Pump

There is also a certain changing rule between the head of pump and flow, pump power is the power applied to the pump shaft by the impeller. In order to distinguish the difference between pump power and motor power, the power of the pump is usually called shaft power. In normal use, the rule of the power and flow for a specified pump is fixed; only with the increase of service years, due to wear and corrosion of the pump

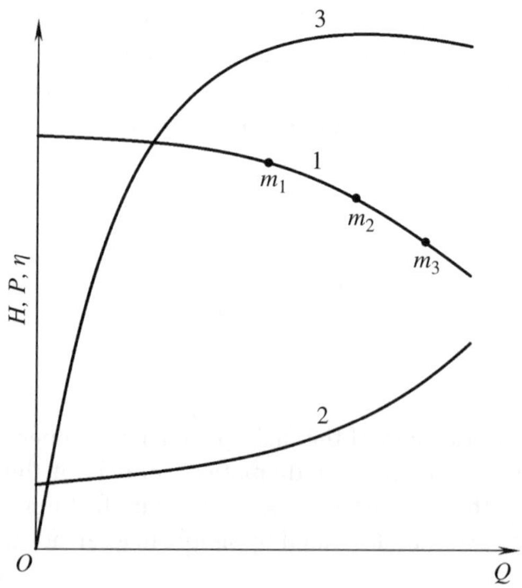


**Figure 5.6** Basic characteristic curve of pump.

impeller, flow channel of water pump housing, bearings and sealing element, this rule will be changed. The power–flow curve of pump is shown in Figure 5.6 (Curve 2). The shaft power of one working point can be obtained from Curve 2, thus the output power of the motor is also obtained.

#### 5.2.2.3 Efficiency–Flow Curve of Pump

The relationship of pump efficiency and flow is shown in Figure 5.6 (Curve 3). The efficiency of one working point can be obtained from Curve 3, which can clearly show whether it is energy saving in this working state and whether it is running efficiently.

Note: Those curves are given in rated speed. At the time of delivery, pump manufacturers should test out the H-Q characteristic curve, *P-Q* characteristic curve, and *η-Q* characteristic curve under the rated speed, and draw them on the pattern to provide to the users. The *H-Q* characteristic curve, *P-Q* characteristic curve, and *η-Q* characteristic curve would change when the pump is in speed control state.

#### 5.2.2.4 Working Point of Pump

The intersection of the *H-Q* characteristic curve of the pump and pipe network characteristic curve is the stable working point of pump. Figure 5.7 shows the intersections of head and flow characteristic curves and pipe network characteristic curves at different speeds. Curve 1 is the *H-Q* characteristic curve at high speed, and the intersection of Curve 1 and pipe network characteristic curve is *A*; Curve 2 is the *H-Q* characteristic curve at low speed, and the intersection of Curve 2 and pipe network characteristic curve is *B*. *A* and *B* are the pump working points at different speed.

### 5.2.3 Theoretical Analysis of Pump Speed Control Energy Saving

According to the similarity law of fluid mechanics, when the speed of water pump changes from $n_1$ to $n_2$, the corresponding relationship of parameters of the pump and speed is as follows:

$$\frac{Q_1}{Q_2} = \frac{n_1}{n_2} \tag{5.15}$$

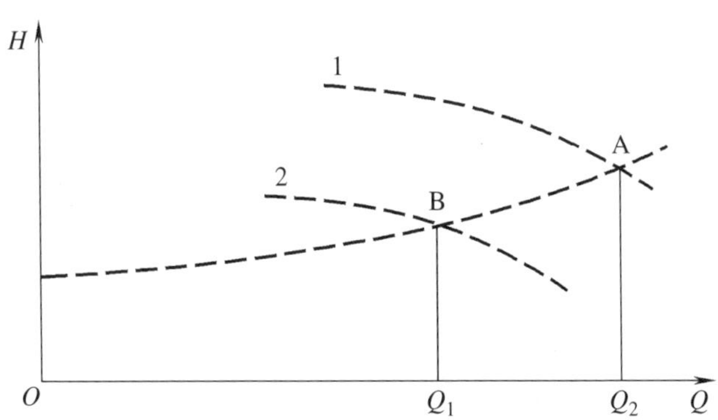


**Figure 5.7** Head and flow characteristic curve of pump.

$$\frac{H_1}{H_2} = \left(\frac{n_1}{n_2}\right)^2 \tag{5.16}$$

$$\frac{P_1}{P_2} = \left(\frac{n_1}{n_2}\right)^3 \tag{5.17}$$

where $Q_1$, $Q_2$, $H_1$, $H_2$, $P_1$, and $P_2$ are the flow, head, and shaft powers at different speeds.

Since the power is proportional to the cube of the speed, the energy-saving effect of variable frequency speed control is quite remarkable in theory. Ignoring the static head, the pipe network characteristic curve equation can be written as

$$H = RQ^2 \tag{5.18}$$

The pump characteristic curve equation at the rated speed is

$$H = a - bQ^2 \tag{5.19}$$

Set the actual flow as $Q_P$, the pump head required by pipe is $H_p$; substituting $Q_p$ and $H_p$ into (5.18), the pipe network resistance coefficient $R$ can be obtained. The coefficients $a$, $b$ in (5.19) are fitted by the data which provided by manufacturers.

Combining (5.18) and (5.19), the pump working point at rated speed ($Q_N$, $H_N$) can be solved out. The pump rated power at rated speed is

$$P_N = \frac{\rho g Q_N H_N}{1000\eta_N} \tag{5.20}$$

Assuming that the flow decreases from $Q_N$ to $Q_p$, if the change of the flow is not large, the flow is regulated by a valve. According to (5.15), when the flow is $Q_p$, the pump speed is

$$n_p = \frac{Q_p}{Q_N} n_N \tag{5.21}$$

The pump shaft power is

$$P_p = \frac{\rho g Q_p H_p}{1000\eta_p} \tag{5.22}$$

where $\rho$, the fluid density;

$\eta_p$, the pump efficiency at working point ($Q_p$, $H_p$). Because the pump efficiencies are approximately equal near the working point, it can be considered that $\eta_p$ is equal to the rated efficiency $\eta_N$.

So the saving power of pump by using speed control strategy is

$$\Delta P = P_N - P_p \tag{5.23}$$

This is the energy-saving value of the pump and pipe network after using speed control strategy. For the speed control system, the loss of the speed control device should also be taken away.

### 5.2.4 Energy-saving Calculation of Variable Frequency Speed Controlling Water Supply System

The constant speed driven pump is analyzed theoretically above; when water demand decreases, using valve to adjust flow will lead to energy loss. The energy saving of variable frequency speed controlling water supply system will be analyzed by an example in the following.

**Example 5.1** The water supply demand of one water station is 120 $m^3$/h, the required head of the pipe provided by the pump is 70 m. Now there is a pump, rated flow 140 $m^3$/h, head 90 m, efficiency 63%, and speed 2950 r/min, the pump speed–flow characteristic equation at rated speed is $H = 102.586 - 0.0006883Q^2$. Using variable frequency speed control to meet the above situation, try to determine the pump speed and calculate the energy-saving effect.

**Solution**

1) First, solve the pump working point.
   At working point $P$, $Q_p = 120 m^3/h = 0.0333 m^3/s$, $H_p = 70 m$, according to (5.18), the pipe network resistance coefficient can be solved out:

$$R = \frac{H_p}{Q_p^2} = 0.00486$$

   So the pipe network characteristic equation is obtained.

$$H = 0.00486Q^2$$

   Combine the above pipe network characteristic equation with the pump characteristic equation at rated speed, the pump working point at rated speed can be worked out, that is, $Q_N = 136 m^3/h$, $H_N = 89.86 m$.
2) The speed corresponding to actual flow.
   According to (5.21), the speed corresponding to actual flow can be solved out.

$$n_p = \frac{Q_p}{Q_N} n_N = \frac{120}{136} \times 2950 = 2603 r/min$$

   The shaft power at speed $n_p$ is

$$P_p = \frac{\rho g Q_p H_p}{1000\eta_p} = \frac{1000 \times 9.8 \times 0.03333 \times 70}{1000 \times 0.63} = 36.29 kW$$

   where $\eta_p = \eta_N = 0.63$.

   If using value to adjust flow, when $Q = 120 m^3/h$, the actual head provided by the pump is

$$H = 102.586 - 0.0006883 \times 120^2 = 92.6 m$$

The pump shaft power is

$$P_N = \frac{\rho g Q H}{1000\eta} = \frac{1000 \times 9.8 \times 0.033 \times 92.6}{1000 \times 0.63} = 47.53\text{kW}$$

where $\eta = \eta_m = 0.63$.

3) The power saved by variable frequency speed control is

$$\Delta P = P_N - P_p = 47.53 - 36.29 = 11.24\text{kW}$$

The energy-saving rate of variable frequency speed control is

$$\Delta E = \frac{\Delta P}{P_N} = \frac{11.24}{47.53} = 23.6\%$$

This is the calculation result by different points of the pump characteristic curve, and the loss of the frequency converter itself should also be taken away.

## 5.3 Control Principle of Constant Pressure Water Supply System

### 5.3.1 Control Principle of Constant Pressure Water Supply

From the running situation of the traditional water supply control system, there exist the following several problems that restrict the quality of water supply: first, the users' water consumption change with seasons, working days, and the differences of day and night, making the water supply difficult to control. However, the whole water supply system makes a larger design margin in order to satisfy the requirement of maximum water supply, which would waste a large amount of energy in low water consumption. Second, operators need to frequently increase or decrease the pump running numbers or regulate valve to maintain the balance between water supply and demand, which greatly increases the workload of operators. This method is difficult to ensure the water pressure constant; it will fluctuate in a certain scope, affecting the quality of water supply. Third, frequently opening the water pump increases the mechanical shock of the motor and pump.

It can be seen that adjusting motor speed by water consumption (flow) can improve the quality of water supply and eliminate the energy loss caused by the surplus flow. But the flow is a physical quantity which is not easy to acquire directly in the actual problem. For example, in Figure 5.2, the flow $Q(t)$ is not easy to be detected or determined directly at a certain moment. So the physical quantities that are easy to be controlled directly should be researched in the water supply system.

According to (5.3) and (5.4), the following can be obtained:

$$Q = Q_0\sqrt{\frac{H}{H_0}} \tag{5.24}$$

Equation (5.24) shows that flow can be controlled by head. When the pump flow remains the same, the head is proportional to the export pressure. Thus, the flow can be controlled by pressure. That is to say, controlling the outlet pressure of water supply system can realize the flow control, and finally meet users' demand for flow. So pressure is the control object of water supply system. Assuming $Q_g$ represents the water supply capacity (flow), $Q_u$ represents the water demand. The pipe network pressure can reflect the relationship of the water supply capacity and the water demand.

If water supply capacity is greater than demand, namely, $Q_g > Q_u$, the pipe network pressure increases.
If water supply capacity is less than demand, namely, $Q_g < Q_u$, the pipe network pressure decreases.
If water supply capacity is equal to demand, namely, $Q_g = Q_u$, the pipe network pressure remains the same.

Therefore, the contradiction between water supply and water demand is reflected on the change of the pipe network pressure, and pressure becomes the reference of flow controlling. Thus, keeping the pipe network pressure as a constant value could guarantee the balance of water supply and water demand and properly meet the users' requirements, which is the basic principle of constant pressure water supply control.

### 5.3.2 Constant Pressure Water Supply Control System

The constant pressure water supply system is realized by a frequency converter and the closed loop of pressure. Figure 5.8 is the basic structure of the constant pressure variable frequency speed control water supply system. In the figure, water source comes from a pipe network. As for water supply of high buildings, the pressure coming from a water pipe network is not enough, so the second pressure boost is provided by the pump. A pressure sensor is installed on the main pump export. The pressure given by the system and the pressure acquired by the sensor are passed on to the controller, and

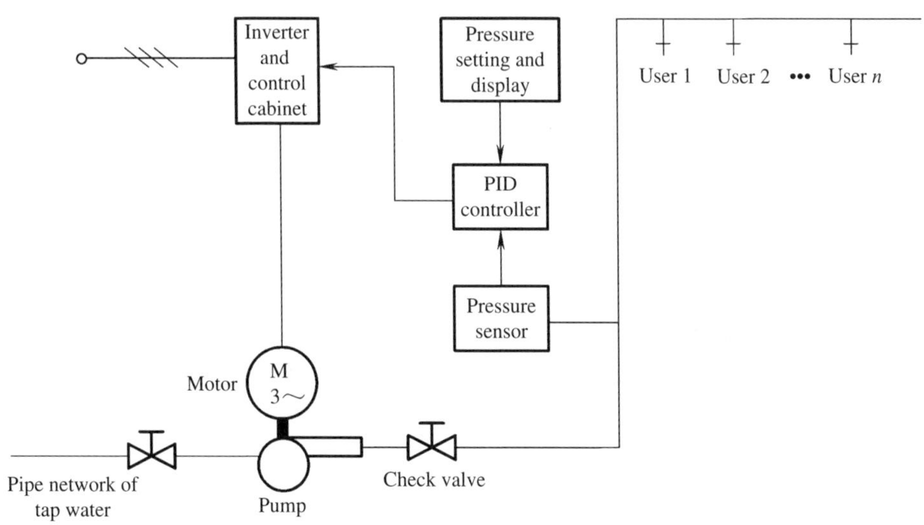


**Figure 5.8** Schematic diagram of constant pressure variable frequency speed control water supply system.

the converter's frequency is calculated by the controller; then the motor's speed is controlled by adjusting the converter's frequency.

In the flow changing water supply system, the equilibrium point of water supply and water demand is codetermined by the system pipe network characteristic curve and the pump characteristic flow–head curve. The intersection of the two curves is the pump operating point; the pump flow just meets users' requirements at that time, and the speed corresponding to this point is exactly the desired speed that is adjusted by the frequency converter. To ensure the constant pressure water supply, the pump operating points need to be adjusted continually to balance the supply and demand of flow; the pressure at that moment is the pressure given by the constant pressure water supply control system, and the pressure at one point detected by the sensor is the feedback quantity. The control of the motor's speed is realized by pressure detection and comparison, which can balance the supply and demand of the water supply system all the time.

The control principle of constant pressure water supply system is shown in Figure 5.9. Supply pressure is passed on to the control system through the pressure sensors: the given pressure of the system is $P_g$, feedback pressure is $P$, and error is $\Delta P$. The converter's frequency is adjusted by a proportional–integral–derivative (PID) controller.

There are two kinds of commonly used input control signals of converter frequency: one is digital signal, and the other is analog signal. The input of digital signal is pulse train, and its frequency is proportional to the frequency of the converter. The input of analog signal is analog voltage or current. Usually there are three kinds of forms: one is 0–5 V, the corresponding frequency is 0–100 Hz; second is 0–10 V, the corresponding frequency is 0–10 0Hz; third is 4–20 mA, the corresponding frequency is 0–100 Hz. The 4–20 mA control signal has strong anti-interference ability, which is beneficial to long-distance transforming. Theoretically, the frequency of the converter begins with zero, but actually starts with near 1 Hz. The upper frequencies of different converters are not the same, some are 100 Hz, 400 Hz, or even higher.

In Figure 5.9, the PID controller has proportional, integral, and differential adjusting functions. When feedback pressure $P$ is less than $P_g$, the output voltage of the PID controller increases; the operating frequency of converter goes up, which increases the speed of motor; and the pressure of pipe network $P$ becomes higher. When feedback pressure $P$ is higher than $P_g$, the output voltage of PID controller decreases; the operating frequency of converter goes down, which decreases the speed of motor; and the pressure of pipe network $P$ becomes lower. Therefore, controlling the speed of the

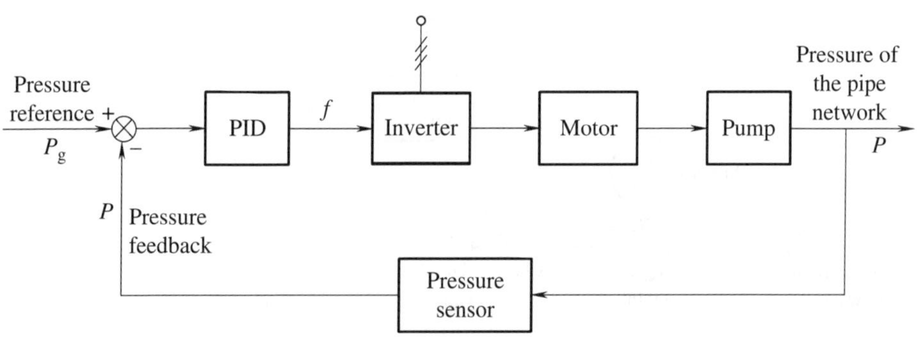


**Figure 5.9** Control principle diagram of constant pressure water supply system.

motor by the pipe network pressure and regulating the flow of water pump can meet the users' requirements commendably.

Meanwhile, in the valley usage of water, the users' demand reduces, the flow of control pump decreases, so the shaft power of pump decreases, compared with the rated flow, the motor power reduces greatly, achieving the purpose of energy saving.

In Figure 5.9, the constant pressure water supply system is achieved by a single-chip microcomputer. The microcomputer compares the collecting pressure signal with the given pressure, calculates PID parameters according to the pressure difference, and the frequency of the converter is given. When the pressure difference is zero, the frequency given by the PID controller corresponds to the given pressure of the pipe network. Sometimes, the programmable logical controller (PLC) can be used in constant pressure water supply control system, because the reliability of PLC is higher than the microcomputer and PLC operation function becomes more and more strong. Some frequency converters contain a built-in PID controller, which provides a more convenient condition to the constant pressure water supply system.

Compared with traditional ways of water supply, constant pressure water supply control system not only has the advantage of energy saving, but it can also truly realize the constant pressure water supply, which can solve the shortcoming that the traditional water supply system is difficult to control in the top or bottom of water usage. Meanwhile control cabinet can perform the function of automatic switching of the pump, which can reduce labor cost. In the design of control system, the switching control of frequency conversion and power frequency should also be considered. In the event of the failure of the frequency converter, the motor automatically switches to the power supply, which can ensure the continuity of users' water supply.

## 5.4 Application of Variable Frequency Speed Control Energy-saving Technology

### 5.4.1 Basic Principle of Motor Variable Frequency Speed Control

It can be seen from all methods of speed control that variable frequency speed control has the highest comprehensive performance index, so variable frequency speed control is extensively used in the actual production. Motor variable frequency speed control device (frequency converter) is a kind of complex power electronic equipment, in order to make better applications of variable frequency speed control technology; the following is a brief introduction of the working principle of variable frequency speed control.

According to the frequency converter topology and control methods, the frequency converter can be divided into AC–AC converter and AC–DC–AC converter.

AC–AC converter turns alternating current (AC) of grid directly into the AC that both the voltage and the frequency are adjustable; since it does not need intermediate direct current (DC) link, it is also called direct frequency converter. For AC–AC converter, there is no intermediate DC link, so it has many advantages, such as higher efficiency, easy four-quadrant operations, and good sinusoidal output waveform in low frequency. But due to the restrictions of the grid frequency and the pulse number of converter circuit, the output frequency is low. Normally, it is only about 1/2–1/3 of grid frequency. It also has some other disadvantages, such as low power factor, large current

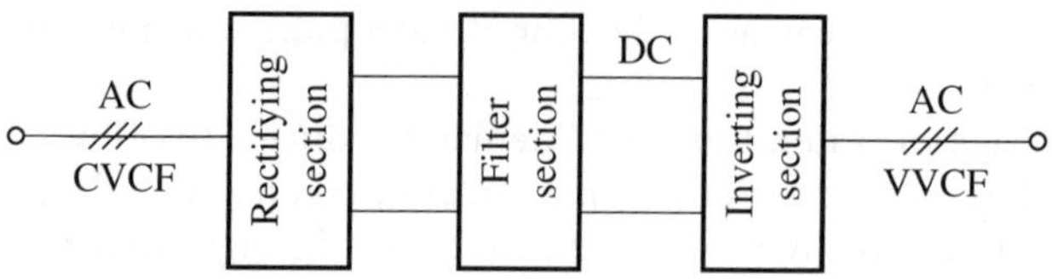


**Figure 5.10** Schematic diagram of AC–DC–AC converter circuit.

harmonic content, and so on. Due to these shortcomings, the applications of AC–AC converter are limited.

AC–DC–AC converter first rectifies AC into DC, then converts DC into AC that the frequency is adjustable, so it is also called indirect frequency converter. The schematic diagram of AC–DC–AC converter circuit is shown in Figure 5.10.

The figure includes a rectifying section, a filter link, and an inverting section. The rectifying section rectifies constant voltage constant frequency (CVCF) AC into DC. Filter link filters the output pulse waveform of the rectifier into a constant DC, providing reactive energy storage for the inverter section. The inverting section plays the role of inverting DC into AC of variable voltage variable frequency (VVVF). Filter can be a capacitor or an inductor (reactor), corresponding to voltage converter and current converter, respectively. The so-called voltage converter is to maintain a constant DC link voltage, namely, the DC side voltage has no abrupt change. The current converter is to maintain a constant current, that is, the current through the inductor does not have a sudden change. Voltage converter is the most widely used type of the converters.

The control methods of frequency converter are based on PWM technology at present, which modulates the output pulse width to equally obtain the sinusoidal output waveform. Each pulse width and interval of PWM waveform can be accurately calculated. It can also be achieved by using the modulation method that treats the expected output waveform as the modulated signal and the modulated signal as the carrier wave; the desired PWM waveform can be obtained by the modulation of signal waveform. We usually use triangular wave or saw-tooth wave as the carrier wave, sine wave as modulation signal, and the waveform obtained is SPWM waveform. This carrier-frequency triangular wave comparison method has been widely used in the actual control.

Vector control technology has been extensively used in recent years. From the working principle of an asynchronous motor, we can get a spatial magnetic field rotating with power frequency by providing a symmetrical three-phase sinusoidal current $i_A$, $i_B$, $i_C$ to the spatially symmetrical three-phase stator windings U, V, W. In another way, we can also get a spatial magnetic field by providing two-phase sinusoidal AC current $i_\alpha$, $i_\beta$ whose mutual time difference is 90° to the two-phase stator windings $\alpha$, $\beta$ whose mutual space difference is 90°. Providing DC current $i_M$, $i_T$ to perpendicular static winding also would generate a flux $\dot{\Phi}$ which is fixed to winding in spatial position. Also flux will rotate with the speed of $\omega_1$ in space if winding rotates with the speed of $\omega_1$. From the point of producing a rotating field, the three sets of windings above is completely equivalent.

Abstracting windings M, T as a two-phase rotating coordinate system M-T, and taking the magnetic flux $\dot{\Phi}$ in the direction coincident with the M-axis, when the observer analyzes from M-T coordinates, the flux is generated by $i_M$ alone, and $i_T$ is only used to generate electromagnetic torque, so the asynchronous motor is equivalent to a DC motor, M winding equivalents to the excitation winding, and T winding equivalents to the armature winding. Since this equivalence needs the help of a current space vector (on behalf of magnetomotive force) coordinate transformation, we describe this transformation that the control system realized by the coordinate transformation as the Vector Control System. After coordinate transformation, a three-phase asynchronous motor is converted into an equivalent DC motor. By imitating the control strategies of a DC motor, we can control a three-phase asynchronous motor with the control methods of DC motor and obtain a good speed performance.

Vector control technology makes it possible to get good static and dynamic performances in AC speed control system as DC drive. However, as the rotor flux is difficult to be observed accurately, the system characteristics are greatly influenced by motor parameters and the complexity of the vector rotational transform, the actual control effect is difficult to achieve the results of the theoretical analysis in practical application.

The control method of Direct Torque Control (DTC) is to analyze the mathematical model of the AC motor in the stator coordinate system, calculate the motor's flux linkage and torque, and compare torque detecting value with the desired value by using the discrete two-point regulator (Band-Band Control), so that the torque fluctuation is limited within a certain tolerance range. Therefore, it would generate a PWM pulse width modulation signal to control the switching state of the inverter directly, which can get a high dynamic performance of the output torque. Therefore, DTC adjusts the relative position of the stator flux $\Psi_s$ and rotor flux $\Psi_r$ to obtain torque characteristics and speed performance with high dynamic response, and this is the basic control strategy of DTC, which has been widely used in the high-performance frequency converter currently.

## 5.4.2 Selection of Frequency Converter

The correct choice of a frequency converter is very essential not only for meeting the technological requirements, but also for realizing the energy saving of the motor drive system. For example, for a quadratic torque load, universal converter of U/F control mode should be chosen; for a constant torque load (lift or crane), the constant torque converter should be chosen. If making an improper choice, the energy-saving effect would be greatly reduced, or it would not achieve the effect of energy saving; what is even worse is that it may destroy frequency converter or motor. In addition, the prices for different types of converters are not the same.

### 5.4.2.1 Type Selection of Converter

The selection of frequency converter types should be in accordance with the requirements of the load characteristics. For a general load that only needs simple speed control, such as fans, pumps, woodworking machinery, conveyors, mixers, general machine tools, and extruders, the general U/F control mode universal converter can satisfy the desire. For the high-precision machine tool spindle drive, elevators, cranes, lifts, steel industry, high-speed papermaking, packaging, printing machinery, and other high-speed high-precision control occasions, high-performance vector control type (or DTC type) converter is necessary.

#### 5.4.2.2 Power Supply Selection of Converter

According to the different national industrial standards or purposes, the voltage levels are not identical. When choosing frequency converters, we should note that whether the voltage level, power frequency, and supply voltage range are compatible with the voltage level of the input power and the drive motor. The voltage level of the low voltage general frequency converter is divided into 220 and 380 V in China, and 690 V for mine, 6000 V, 10,000 V for the speed control of high voltage motor. Voltage, frequency, and the number of phases, as well as the range of voltage and frequency are the major requirements for frequency converters. The phase of the power refers to single-phase or three-phase. Usually the voltage range is about 200–230 V for single-phase and 380–460 V for three-phase. There are two kinds of frequency: 50 and 60 Hz.

#### 5.4.2.3 Frequency Characteristic Selection of Converter

The technical requirements of the output frequency range of converter, output frequency resolution, output frequency accuracy, frequency setting mode (analog, digital, and communication), frequency setting resolution, and U/F characteristics are proposed according to the production process. As for the vector control converter, the speed control accuracy, torque, and the dynamic response time are also considered as the basic technical requirements of the selection of frequency converter.

#### 5.4.2.4 Function Selection of Converter

The choice of converter function is based on the production process requirements. General converter functions are U/F curve setting, acceleration and deceleration time settings, panel operation running, external terminal operation running, communication control, and so forth. In addition, there are also some dedicated functions such as built-in multispeed control, simple PLC, process PID, traverse function for textile and torque control for constant tension control system, and so forth.

#### 5.4.2.5 Capacity Selection of Converter

The selection of converter capacity is determined by many factors, such as motor capacity, the rated current of motor, acceleration and deceleration time, overload multiples, and so forth.

For a converter driving a motor case, the major task is to make sure that the rated capacity of the motor (apparent power) is less than the rated output capacity of the converter. There are two algorithms of the rated capacity of the motor, namely

$$\frac{k_{\mathrm{ic}} P_{\mathrm{N}}}{\eta_{\mathrm{N}} \cos \varphi_{\mathrm{N}}} \leq S_{\mathrm{in}} \tag{5.25}$$

$$k_{\mathrm{ic}} \sqrt{3} U_{\mathrm{N}} I_{\mathrm{N}} \times 10^{-3} \leq S_{\mathrm{in}} \tag{5.26}$$

At the same time, the following requirements for current should be met:

$$k_{\mathrm{ic}} I_{\mathrm{N}} \leq I_{\mathrm{in}} \tag{5.27}$$

where $S_{\mathrm{in}}$, the rated output capacity of converter (kVA);
$I_{\mathrm{in}}$, the rated output current of converter (A);

$P_N$, the rated power of motor (kW);
$U_N$, the rated voltage of motor (V);
$I_N$, the rated current of motor (A);
$\eta_N$, the rated efficiency of motor;
$\cos\varphi_N$, the power factor of the motor in the rated operation;
$k_{ic}$, the compensation coefficient of the current waveform, which is generally 1.05–1.1 as the converter output current waveform contains higher harmonic components.

In addition, for the motors of the same power and different poles, the currents are different. The more poles the motor has, the greater the current is. The rated capacity of motor should be less than the rated output capacity of the converter in both calculation methods.

Please refer to other relevant information for the case of a converter driving several motors.

#### 5.4.2.6 Selection of Other Accessories

After the above selection, requirements for the converter have been proposed clearly. Other accessories must be decided whether to be selected by the requirements, such as the braking unit, feedback unit, input reactor, output reactor, DC reactor, and so on. For example, in the frequency transformation of gantry planer and milling machine, due to the need of quick go and back action, it has to select DC braking unit or feedback unit as well as high-performance vector control converter. As for hospital elevator converter, in order to reduce the interference of the converter to the grid and to prevent interference to medical equipment, an input reactor or DC reactor should be added in general.

### 5.4.3 Instances of Converter Selection

**Example 5.2** A centrifugal fan model is 4-72-11208, its flow is 227,500 $m^3$/h at the rated speed, full pressure is 2400 kPa, and spindle speed is 720 r/min. The equipped asynchronous motor model is Y355L-8, rated power is 200 kW, rated efficiency is 94%, rated speed is 730 r/min, frequency is 50 Hz, rated voltage is 380 V, rated current is 390 A, and the rated power factor is 0.83. The fan is used for ventilation of production workshop, long-term working. The air volume is adjusted by adjusting the degree of the inlet dampers' opening originally. In the short-time blowing case, the shutter opening degree is 100%, and the fan outlet velocity is 6 m/s. Under normal circumstances, the shutter opening degree is 60% and the fan outlet velocity is 4 m/s, satisfying the requirements of the air volume in the normal production. Try to select the motor and the frequency converter that match the fan.

**Solution**

Usually adjusting the wind speed and flow by the baffle would cause energy wastage. Now it uses frequency conversion velocity modulation instead of constant speed motor driving in order to save energy. The steps according to the above situation that are used to choose converters are as follows:

1) Type Selection of Converter: The load of motor is centrifugal fan, so constant U/f control mode converter is selected.
2) Voltage Selection of Converter: The rated voltage of motor is 380 V, so 380 V is selected as the converter voltage.

3) Frequency Characteristic Selection of Converter: The speed of the motor is adjusted downward from the rated speed, so there is no special requirement for the frequency, and the general converter with basic functions can meet the requirements.
4) Function Selection of Converter: As the torque is proportional to the square of the fan's speed, and the motor power is large, so current type converter is available.
5) Capacity Selection of Converter: In practical applications, the shaft power of motor should be calibrated first, and the capacity of the converter cannot be determined by the power on motor nameplate simply.

Calculate the flow of the fan according to the basic parameters of the fan features.

$$Q = \frac{Q_N}{3600} = 63.2\text{m}^3/\text{s}$$

The shaft power of fan is

$$P_{W1} = \frac{QH}{\eta_W} \times 10^{-3} \tag{5.28}$$

where $P_{w1}$, the shaft power of fan (kW);
$Q$, the flow of fan ($\text{m}^3$/s);
$H$, the pressure of wind (kPa);
$\eta_W$, the efficiency of fan.

Find out that $\eta_W = 0.8$; substituting the known parameters into Equation (5.28), the fan shaft power can be obtained

$$P_{W1} = \frac{63.2 \times 2400}{0.8} \times 10^{-3} = 190\text{kW}$$

Consider that the rated speed of the fan is 720 r/min, while the rated speed of the equipped motor is 730 r/min. The fan shaft power is proportional to the cube of the speed. Calculate the shaft power of the fan at 730 r/min.

$$P_{W2} = P_{W1}\left(\frac{n_2}{n_1}\right)^3 = 190\left(\frac{730}{720}\right)^3 = 198\text{kW}$$

Fan is connected to the motor shaft directly, the power of motor is 200 kW, and the fan shaft power matches the motor power.

According to Equation (5.25), calculate the capacity of the converter

$$S_{in} = \frac{k_{ic}P_N}{\eta_N \cos\varphi_N} = 1.1 \times \frac{200}{0.94 \times 0.83} = 282\text{kVA}$$

The output current of the converter is

$$I_{in} = \frac{282000}{\sqrt{3}380} = 428.5\text{A}$$

Substituting into (5.27), the rated current of motor is 390 A

$$I_{\text{in}} \approx 1.1 \times 390\,\text{A}$$

So the capacity of converter should be greater than 282 kVA, the selection should be not less than 300 kVA.

For the load of fan (or pump), selecting a general converter can meet the requirements. But note that the torque load of the fan is proportional to the square of the speed, namely, proportional to the square of the frequency of the motor supply. For the motor, the output torque is

$$T_2 = C_{\text{m}} \phi_{\text{m}} I_2 \cos \varphi_2 \tag{5.29}$$

In the equation, $I_2$ is the rotor current, and its value should be less than or equal to the rated current of motor because of the limit of winding fever, so the torque of the motor is proportional to magnetic flux $\phi_{\text{m}}$ when $I_2$ is constant, while $\phi_{\text{m}}$ is proportional to motor terminal voltage when ignoring the stator voltage drop. In order to ensure that the motor output torque and the load torque are equal, the terminal voltage of the motor should be proportional to the square of the frequency. Thus the motor can run in the rated current, which makes torque meet the load requirements in high frequency.

Frequency converter output voltage–frequency characteristics required by the pump load is shown in Figure 5.11 (OAB), which means U/F should meet the law of quadratic curve. If the voltage–frequency characteristic of constant torque was chosen instead (Figure 5.11 OCB), the motor would not run in the best condition; therefore, the motor reactive current increases and the power factor declines.

### 5.4.4 Points Requiring Attention in the Operation of Converter

#### 5.4.4.1 Harmonic Problems

Since the input current of the converter contains a large amount of harmonics, with the extensive use of converter in industrial field, harmonic problems have become more

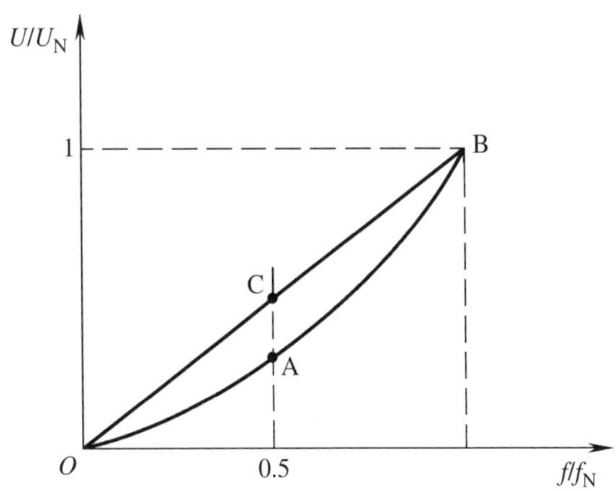


**Figure 5.11** Output frequency–voltage characteristics of converter.

and more prominent, and the pollution of grid has been even worse. The adverse factors caused by current harmonics are mainly reflected in the following aspects:

1) The harmonic component of the current will cause the motor to generate an additional torque, so that the motor output torque decreases, causing the torque ripple.
2) The carrier frequency component in the harmonic current will cause the vibration of the core of the motor and generate electromagnetic noise.
3) The harmonic current of carrier frequency components has a strong radiation ability, which may affect the normal operation of the surrounding radio, computers, and other electronic equipment. Therefore, when using the converter, the rational harmonic suppression measures should be chosen according to the site conditions, such as choosing a DC reactor, input AC reactor, or electromagnetic interference filters. First, make sure the converter does not interfere with its own converter control equipment, and try to minimize the pollution to the power grid. In addition, converter manufacturers are also trying to develop grid pollution-free green frequency converters to solve the harmonic pollution problems from the source.

#### 5.4.4.2 Torque Ripple Problems

The use of a frequency converter has solved the issue of production process and improved the production efficiency. However, due to the requirements for turn-on time and turn-off time of the power devices, generating the dead zone of PWM would cause the distortion of current at the zero crossing point. The distortion current together with harmonic voltage caused by PWM would make the output torque ripple, particularly prominent in the low-frequency operation. As for the production machinery with stringent requirements for torque ripple, it is recommended that the minimum operating frequency is improved by importing the mechanical reduction device.

#### 5.4.4.3 Interference Problems

The harmonics of the input/output circuits would generate a huge radiation interference when using the frequency converters, and could cause the control system malfunction or even could not run in severe cases. Taking these issues into consideration, when using converters in control system, attention must be paid to the wiring of the system, as for the power of control system, a separate EMI filter can be added to improve the anti-interference capability. In addition, when there is a long connecting wire between the converter output and the motor, radiation interference can be reduced by adding an output AC reactor; at the same time, reduce the leakage current of the output line caused by $dv/dt$, extending the wiring distance. A shielded cable can be used to improve reliability when necessary.

### 5.4.5 Application of VVVF Energy-saving Technology

The VVVF energy-saving technology is widely used in all walks of life. Besides the water supply system, VVVF energy-saving technology has made a good energy-saving effect in large oil refineries, chemical plants, elevators, central air conditioning, tunnel ventilation, mine ventilation, and so forth. The following describes the application of energy saving in a fan ventilation system and an air compressor.

#### 5.4.5.1 Application of Fan VVVF

The fan has been used in the central air-conditioning, steel mills, cement plants, chemical plants, mining, and other industries. The greatest feature of the fan is that the load torque is proportional to the square of the speed, and the shaft power is proportional to the cube of the speed. Frequency control of the fan directly affects the effectiveness of ventilation, fan life, and power consumption. Therefore, adjusting the motor speed according to the need of air volume (or pressure) can save a lot of energy. When the fan speed is constant, the output power of the motor is invariable, and at this time, the fan opening is usually 50–80%, and the wind deflector would cause a waste of surplus wind. Frequency converter directly controlling air flow is a good solution to this problem. Motor variable frequency speed control is a stepless speed control, allowing the fan to have a smooth start and stop, and more reliable and better operating performance. The following illustrates the energy-saving effect by a frequency control instance of a fan.

**Example 5.3** A chemical plant has an 85 kW fan; a matching motor model is Y280M-4, 90 kW. It runs 330 days a year in average; the opening of the fan is usually less than 70%, leading to a serious waste of energy. Now retrofit its energy-saving technology, investment is about 100,000 yuan, please analyze the economic benefits of the energy-saving transformation.

**Solution**

1) The measured data before the transformation

$$U = 380\text{V},\ I = 160\text{A},\ \cos\varphi = 0.887,\ P = 93.4\text{kW}$$

Annual electricity consumption before the transformation

$$W_{\text{pq}} = P \times 24 \times 330 = 93.4 \times 24 \times 330 = 739{,}728\text{kWh}$$

2) The measured data after the transformation

$$U = 380\text{V},\ I = 100\text{A},\ \cos\varphi \approx 1,\ P' = 65.4\text{kW}$$

Annual electricity consumption after the transformation

$$W_{\text{ph}} = P' \times 24 \times 330 = 65.4 \times 24 \times 330 = 517{,}968\text{kWh}$$

3) An annual electricity saving

$$\Delta W = W_{\text{pq}} - W_{\text{ph}} = 739{,}728 - 517{,}968 = 221{,}760\text{kWh}$$

4) Saving rate

$$\Delta E = \frac{W_{\text{pq}} - W_{\text{ph}}}{W_{\text{pq}}} = \frac{739{,}728 - 517{,}968}{739{,}728} \times 100\% \approx 30\%$$

The electricity price is 0.6 yuan/kWh, the annual saving electricity bill after the technological transformation is

$$221{,}760 \times 0.6 = 133{,}056\,\text{yuan}$$

The energy-saving effect of technological transformation for this wind turbine is very obvious. Full investment will be taken back in around 9 months. On the other hand, using the frequency converter would make the energy-saving transformation of fans simple in structure, and easy in transformation; after using a frequency converter, the fan can realize soft start and stop, which can reduce the mechanical impact of equipment, extend equipment life, and reduce equipment maintenance costs.

#### 5.4.5.2 Applications of Air Compressor VVVF

Air compressor is very common in the application of mining enterprises. The main control of the compressor is air pressure, and in the refrigeration or cooling system, the temperature is usually set as control variables. The capacity of the motor which matched with the compressor is generally larger and mostly run continuously all year round, so there is great potential for energy saving. The traditional compressor control methods are (i) adjusting the valve opening of the input or output port manually, (ii) using machine to unload and load automatically, (iii) changing the angle of the blade to adjust the pressure or flow.

The possibility that the compressor runs at full load cannot be excluded, so the motor capacity is selected according to the maximum demand; thus the design capacity is generally larger. In actual operation, the time in light load operation is more. Adjusting the degree of the valve opening or changing the blade angle cannot reduce the efficiency of the motor, even in the case of a small demand. If adapting the frequency control, the operation efficiency can be greatly improved, which has a great potential of energy saving.

**Example 5.4** A plant has an air compressor of 50 kW; the equipped motor is Y250M-4, 55 kW, 380 V. The frequency converter power should be greater than 1.15 times of the motor power; according to the power level, we select the frequency converter whose power is 75 kVA, the input current is 150 A, so the final choice is POWTRAN PI97G converter. The frequency converter is suitable for a power supply of 380 V, 50–60 Hz, the range of output frequency is 0.5–400 Hz, and it could display the motor set frequency, operating frequency, speed, current, voltage, and so forth. Under the normal

**Table 5.2** Comparison table of power consumption.

| | Before installing the converter | | After installing the converter | |
|---|---|---|---|---|
| **Parameters** | **Pressure** | **Pressure relief** | **Pressure** | **Pressure relief** |
| Wind pressure (MPa) | 0.65 | 0.50 | 0.65 | 0.50 |
| Current (A) | 101 | 98 | 78 | 65 |
| Voltage (V) | 386 | 386 | 388 | 388 |
| Power factor | 0.82 | 0.80 | 0.996 | 0.995 |
| Input power (kW) | 55.4 | 52.4 | 52.2 | 43.5 |

use of ventilation system, the frequency converter tests data before and after the installation is shown in Table 5.2. The voltage, current, and power factor after the installation in the table are the test results of the converter input ports. Air compressor runs 20 h a day on average, pressure and pressure relief time is about half of the working time. It runs 300 days annually. Transform the energy-saving system of this air compressor, the total investment including the frequency converter and control system costs about 50,000 yuan. Try to analyze how long the input costs can be taken back.

**Solution**

1) First calculate out the direct economic benefits
According to Table 5.2, first calculate out the daily electricity savings

$$(55.4-52.2)\times 10+(52.4-43.5)\times 10=121\text{kWh}$$

Electricity price is 0.60 yuan/kWh, daily electricity savings is 72.6 yuan. Annual electricity savings is up to 21,780 yuan.
2) Then, calculate out the indirect economic benefits

After installing frequency converter and control system on the air compressor, the energy-saving effect is notable, and the saving of lubricating oil and decrease of noise are both obvious at the same time. Following is the reason why it can save lubricating oil. When the converter is in pressure relief process, the speed is reduced and the oiler's fuel injection is only 60.5% of the normal speed. In this condition, the oil pressure of the transmission mechanism supplied by the pump declines more than 0.1 MPa and the amount of oil supply drops, 1/3. Considering the saved fuel by extending the oil change time, the total fuel consumption is only 50% of that not using the frequency converter. In addition, in the unloading state, the noise drops from 74.5 to 69 dB; it is thus clear that as mechanical wear decreases, the maintenance workload and accessories' consumption reduce too.

Annual savings of lubricants and maintenance is about 20,000 yuan. Two charges together, it can save up to 42,000 yuan a year. The full investment can be recovered in around 14 months. It is visible that the frequency control has good promotion prospects in terms of air compressors.

## 5.5 Principles of Motor's Pole Changing Speed Control

### 5.5.1 Pole Changing Working Principle of Motor

The so-called pole changing speed control is obtained by changing the combination (connection) way of the stator windings to change the number of poles of the motor, thus to change the motor's speed. As can be seen from Table 5.1, the biggest drawback of pole changing speed control is that the speed differential is so large that it could not adjust speed smoothly. But in some cases, the load only needs 2–3 fixed speeds, and this 2–3 fixed speeds are very close to a motor's pole changing speed, or the speed after the speed reducer corresponds to the pole changing speed of motor. Then pole changing speed control methods can be used to achieve speed matching, and achieve the purpose of energy saving of the motor drive system with a better energy-saving effect.

From the speed equation of a three-phase asynchronous motor

$$n = \frac{60f(1-s)}{p} \tag{5.30}$$

It can be known that changing the number of poles of the stator windings $p$ can change the synchronous speed of the rotating magnetic field $n_0$. When the power frequency is 50 Hz, the number of poles $p = 1, 2, 3, \ldots$, so the different synchronous speeds of $n_0 = 3000, 1500, 1000$ r/min can be obtained. Since $p$ can only be a positive integer, the pole changing speed control is a step speed control, and the differential is relatively large.

According to the working principle of the motor, the variation of power frequency for a cycle is equivalent to the space magnetic field rotating a pair NS pole. For the squirrel cage motors of a pair pole, three-phase windings in 360° space form one pair of NS poles, so the angle of the magnetic field moving in the space when the frequency changes one cycle (electrical angle 360°) also happens to be 360° (mechanical angle 360°). If three-phase windings in 360° space form two pairs of NS poles, the angle of rotating magnetic field moving one pair NS pole in space, only accounts for 180° space mechanical angle when the power changes one cycle, which means the speed slows down in a half, thus the synchronous speed of rotating magnetic field $n_0$ is inversely proportional to the pole pairs $p$. When the asynchronous motor runs under pole changing speed control, and only in the condition that the pole pairs of the rotor are equal to the stator, the average electromagnetic torque could be formed. For wound rotor asynchronous motor, when changing the connection of stator windings to change the number of poles, the connection mode of the rotor windings must be changed as well to maintain an equal number of poles of the stator and rotor, which makes the wiring and controlling become very complicated. When the number of stator poles of the squirrel cage asynchronous motor changes, the number of rotor poles can follow the number of stator poles automatically. Therefore, pole changing speed control is only used for a squirrel cage three-phase asynchronous motor. The key of pole changing speed control is to change the winding connection mode; the following describes the working principle of pole changing windings.

Assume that each phase winding of the motor stator shown in Figure 5.12 is composed of two perfectly symmetrical "half -phase winding." U-phase, for example, is formed by "half-phase winding" $a_1x_1$ and "half-phase winding" $a_2x_2$. When the head and tail of these two "half-phase windings" are connected in series (called series connected), shown in Figure 5.12a, it would form a magnetic field of $2p = 4$ poles, shown in Figure 5.12b. If the head and tail of these two "half-phase windings" are connected in parallel (called anti-parallel), shown in Figure 5.12c, it would form a magnetic field of $2p = 2$ poles, shown in Figure 5.12d. Seen from Figure 5.12b and 5.12d, once the current of any one of the two "half-phase winding" reverses, the number of poles would be doubled (two half-phase windings are connected in series) or reduced by half (two half-phase windings are connected in parallel). This is the pole changing principle of a single winding double poles, such as 2/4 pole, 4/8 pole, and so on. In addition to the simplest and most commonly used methods of double pole, the pole changing goal of non-double poles can also be attained by changing the winding connection, such as 4/6 pole, and so on. There are two common double-pole schemes, one is Y → Y double pole, and the other is the Δ → Y double pole. Three-phase leading-out terminals of the motor of the two pole changing schemes are few with simple wiring and easy control features.

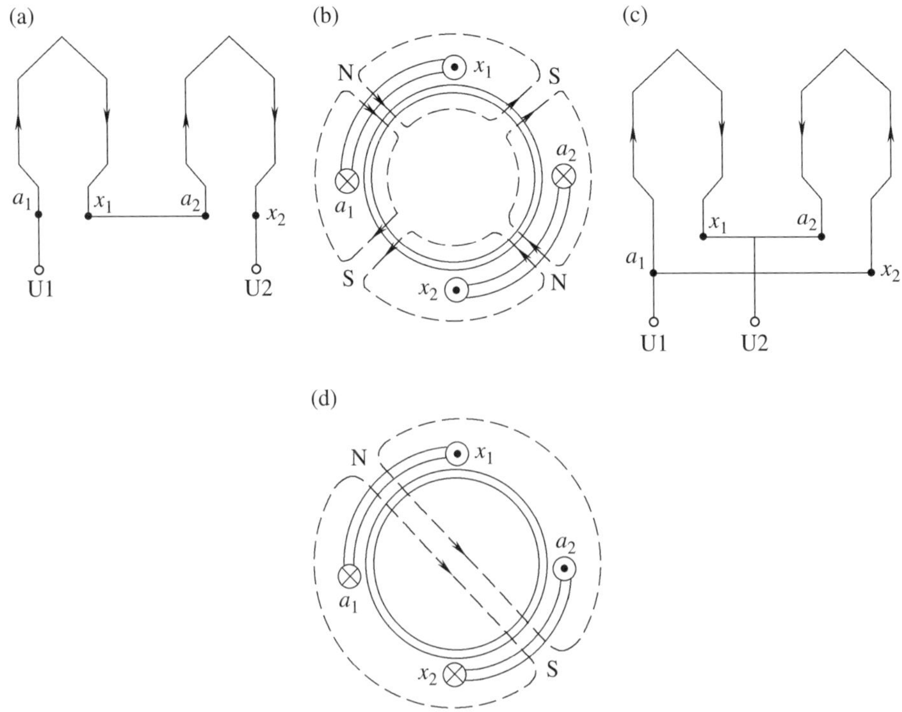


**Figure 5.12** (a)–(d) Wiring diagram of pole changing windings.

It should be noted that in Figure 5.12, in the process of changing the connection of the "half-phase winding," the rotating direction of the rotating magnetic field would also be changed after the change of the pole pairs; in order to maintain constant steering, the phase sequence connection of the power supply must be changed at the same time.

Comparing to the variable frequency speed control, the advantages of the motor's pole changing speed control are high reliability, no more complex speed control equipment, easy maintenance; less investment; no harmonic pollution to the grid; and no supplementary loss to the speed control device itself. The converter itself has about 5–10% loss. Compared to the variable frequency speed control, the biggest drawback of the motor's pole changing speed control is that it is a step speed regulation. The price of a pole changing motor is slightly higher than average motor with the same engine base, the number of the slots in the stator and rotor must be matched when transforming an old motor.

### 5.5.2 Common Pole Changing Methods of Motor

There are generally three ways to change the number of pole pairs of three-phase asynchronous motor stator windings: (i) single winding, placing a set of windings in the stator slots, just change the combinations of its wirings, different number of pole pairs can be obtained; (ii) double windings, putting two separate windings of different number of poles in the stator slots; (iii) put two separate windings of different number of poles in the stator slots, and each winding wire has different compositions to get different number of pole pairs, for instance three or four types. The first method is more simple and practical. Single winding multispeed asynchronous motor has a variety of pole changing methods: the reverse method, commutation method, and variable pitch method are introduced below.

#### 5.5.2.1 Pole Changing Principle of Reverse Method

First, take the simplest 2/4 pole changing two-speed motor as an example to explain the pole changing principle of reverse method. Figure 5.13 indicates a phase winding of the stator, which has only 2 -pole phase groups with concentrated winding. Figure 5.13a and b are the different connections of the 2-pole phase groups. Figure 5.13a is the 2-pole phase groups connected in series, called the tail–head connection. When current passes through them, a 4-pole magnetic field would be generated in the motor according to the right-hand screw rule. Figure 5.13b is the 2-pole phase groups in opposing series connection, called tail–tail or head–head connection. The others are the same as Figure 5.13a. When they are passed through the same current, a 2-pole magnetic field would be formed in the motor. Thus, in order to multiply the number of stator poles, changing the end connection method of the stator windings to reverse the current direction of each half of the phase winding can achieve this purpose. This is the pole changing principle of the reverse method. Figure 5.12 also belongs to the pole changing principle of reverse method, which is the parallel pole changing form of reverse method.

Using the reverse method, it can not only get double speed motor with double-pole ratio (2/4 poles, 4/8 poles, etc.), but nondouble pole ratio double speed motor (4/6 poles, 6/8 poles, etc.) can be obtained. In the following part, take 4/6 poles two-speed motor as an example to introduce its pole changing principles.

In Figure 5.14a, four groups of coils are in series-opposing connection, so the current of coils 1, 2, 3, and 4 generate a 4-pole magnetic field in the motor air gap. If half-phase windings (coils 3 and 4) are in series-opposing connection, as shown in Figure 5.14b, the current in the coil 4 reverses, meanwhile, a 6-pole magnetic field would be produced in the air gap of the motor, so a 4/6-pole double-speed motor can be obtained.

Reverse method to change the number of pole pairs is the most common way to the single winding, double-speed motor.

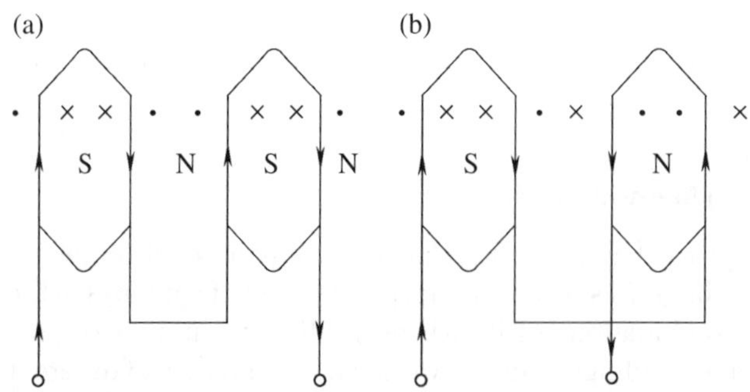


**Figure 5.13** Reverse pole changing method schematic of 2/4 poles. (a) Two poles and (b) 4 poles.

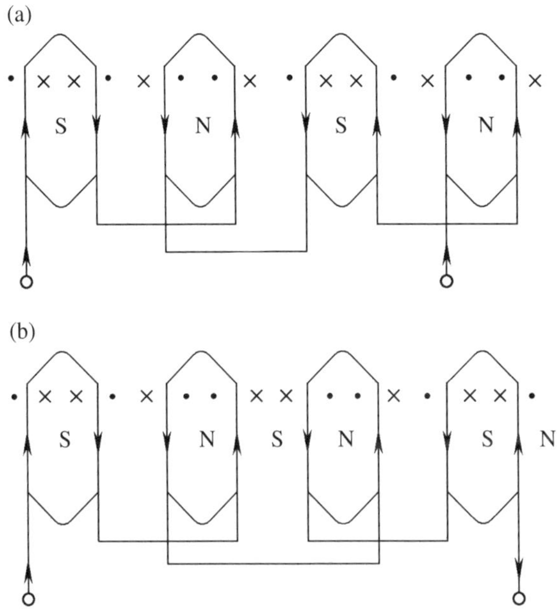


**Figure 5.14** Four-/six-pole changing schematic of reverse method. (a) Four poles and (b) 6 poles.

#### 5.5.2.2 Commutation Method

Reverse method is under the condition of the slot number of motor unchanged, changing the connection to transform the direction of current in the coil, so as to change the number of poles. Commutation method is the reversing part of the coil, at the same time, changing some coils' phase number to achieve the change of the poles. For example, a portion of U-phase coils is changed into a V-phase, a portion of V-phase coils is changed into W-phase, and a portion of W-phase coils is changed into a U-phase. This change will make the stator windings have a higher winding coefficient under certain poles, and the running performance of the motor would be improved under those poles. This makes up for the lack of the reverse method in some windings. The disadvantage of the commutation method is that there are more outgoing ends of the windings. The principle of commutation method and reverse method are similar but different in the ways of changing the wiring connection.

#### 5.5.2.3 Varying Pitch Method

Motor can get two speeds by use of reverse method. Commutation method combined with reverse method can make the motor obtain two, three, or even four kinds of speed. If using the varying pitch method and reverse method, three or more kinds of speed can be gained similarly. The main features of the varying pitch method are that two different pitch windings are embedded in a single winding, one is short pitch winding; other half coil winding is the integral pitch winding, which can meet the need of more pole grades. Changing the connection of the winding can achieve the purpose of changing the number of pole grades. Using this varying pitch method to obtain a single winding three-speed motor, the winding coefficient is high, the outing heads is less, and the control is easy.

### 5.5.3 Common Connection Methods of Wiring Ends

In the above 3 pole changing methods, all are required to change ending connection of the three-phase windings to realize pole changing. In practice, it should not only conform to the needs of pole changing, but also make the connection as easy as possible. For example, in the common connection methods, the outlet terminal of motor with 2Y/Δ, 2Y/Y connection forms has only six outlet heads and 2Δ/Y, 2Y/2Y has nine heads. Figure 5.15 shows the four ending wiring connections of pole changing windings.

In Figure 5.15, the solid line and the broken line indicate the directions of current under two different connections of the windings, respectively. If the currents of each phase are in positive direction under Y- and Δ-connection (predetermined: the

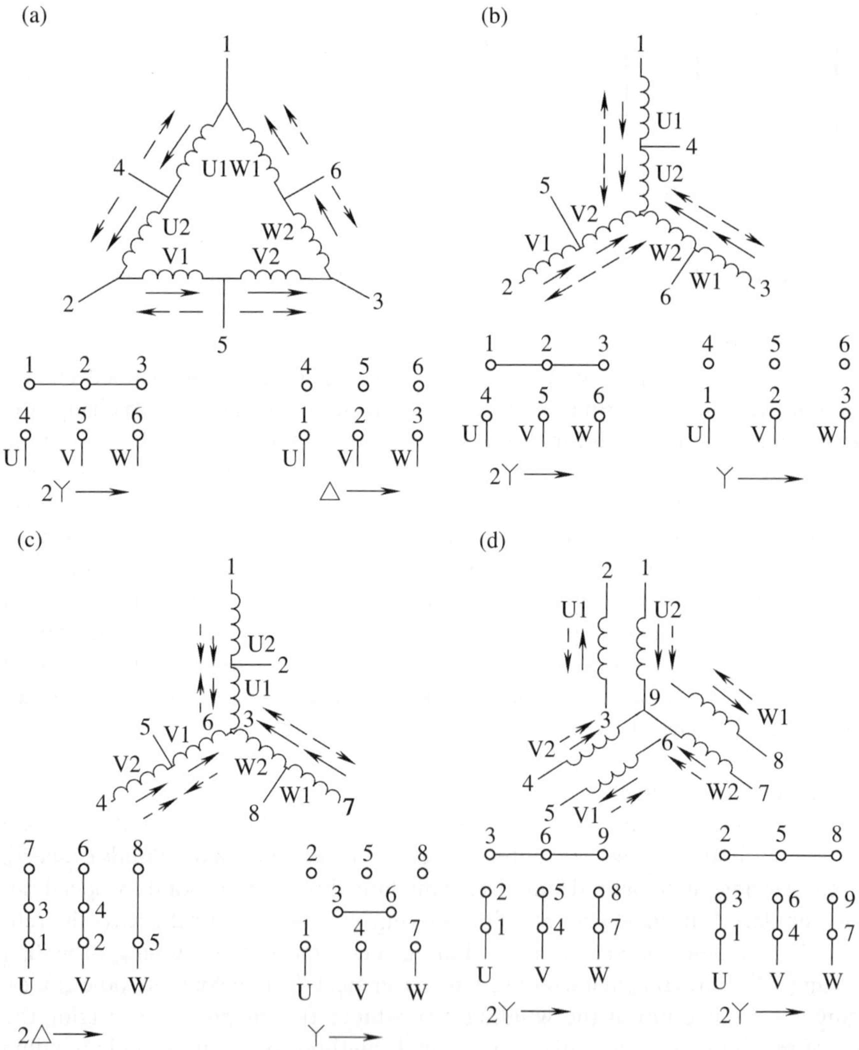


**Figure 5.15** The end wiring schematic of motor's pole changing windings. (a) 2Y/Δ, (b) Y/2Y, (c) 2Δ/Y, and (d) 2Y/2Y.

positive direction of Y-connection current is flowing into the midpoint; the positive direction of Δ-connection current is counterclockwise), then U1, U2, V1, V2, W1, and W2 denote half coils of the U-phase, V-phase, and W-phase windings. As can be seen from the figure, in different connections, currents of U1, V1, and W1 are reversed, meeting the requirement that the currents of each phase need to be reserved when pole changed. In Figure 5.15, the reversal of the current in 2Δ/Y-connection coil is not obvious. If the connections of Y and 2Δ are drawn, respectively, it can be clearly seen that the current directions of U1, V1, and W1 windings become reserved when transforming 2Δ-connection to Y-connection.

Choosing the kind of pole changing connections required changes not only the speed (number of pole pairs), but also the following two points: one is the performance requirements before and after the changes of the motor speed. For example, for 2/4-pole two-speed motors, if using 2Y/Y connection, the output torque of the motor is basically unchanged at the two different speeds, while the output power varies widely, so it belongs to the constant-torque speed control characteristic. If using 2Y/Δ connection, the output powers of the motors are approximately the same at two different speeds, it is the constant-power speed control characteristic. Second is the speed directions of the motor remains the same or not before and after the change. For example, for a 2/4-pole changing two-speed motor, using 2Y/Δ or 2Y/Y connection, the two directions of the motor are reversed before and after the pole changing, while the two directions of some connections keep the same. The above two points must be taken into account when choosing the motor's end connections. Otherwise, the motor would not meet the load requirements.

## 5.6 Energy-saving Principles and Applications of Combined Pole Changing Speed Control

For a single-pump water system (as shown in Figure 5.2), when the flow demand changes, if using the pole changing speed control method to regulate the flow, it is extremely difficult to meet users' requirements. In the system shown in Figure 5.2, the motor has 2 poles, and if the windings are 2/4 poles, the speed of the motor would be decreased from 2900 to 1460 r/min, the speed drop is too large, and it is often difficult to meet the users' requirements. But for multipump systems, under certain circumstances, the use of two (or more) motors combined pole changing speed control method can get close to the effect of variable frequency speed control.

### 5.6.1 Examples of Multipump System

Water injection pump station in oil field is a typical multipump system. In the middle and later periods of oil exploration, reservoir pressure drops, pumping the crude oil to the surface is more and more difficult. In order to maintain reservoir pressure, it needs to inject water to reservoir. Water-injection is essential in the process of oil field production and electricity consumption of water-injection accounts for about 30% of the total electricity consumption of oil production. As the water injection rate is changing in the production process, the water injection motor needs to be under speed control operation. If there is no variable frequency speed control devices in motors, the system

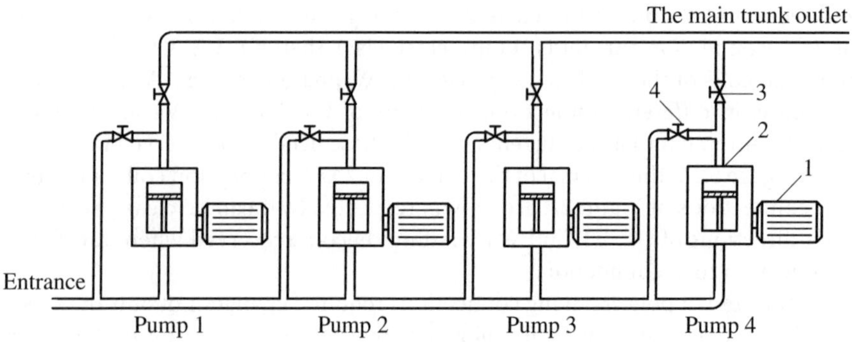


**Figure 5.16** Schematic diagram of multipump injection system.

runs under power frequency constant-speed condition and required to meet the requirements of conditions of water injection, then the water injection volume and pressure are adjusted by the way of back-flow usually, which results in a waste of energy. The following will analyze the reasons of the energy wastage caused by the back-flow through an example.

A pumping station has four injection pumps, and the injection system is shown in Figure 5.16. In the figure, 1 is an electric motor, 2 is a plunger pump, 3 is the check value, and 4 is a return valve. Plunger pumps with high pressure and high-efficiency features are commonly used in the oil field. In the figure, all the water injection pumps are plunger pumps, its rated speed is 1450 r/min, the output volume at rated speed is 65 $m^3$/h, the outlet pressure of plunger pumps is 5 MPa, the power at rated speed is 105 kW, and the rated efficiency of plunger pumps is 82.5%. The power of four motors is 110 kW, rated speed is 1460 r/min, and the rated efficiency of motor is 93.5%.

Normally, the water injection volume is 100 $m^3$/h, the outlet pressure of the main line is 4–5 MPa. The output volume of one pump is not enough, and opening two pumps has 30 $m^3$/h margins. The motor does not configure the speed control device during the design, when the actual water injection volume is less than the pump output volume, back-flow is applied to meet the working conditions. The following is the calculation of power loss caused by surplus flow (back-flow).

1) Input power of the two electric motors is

$$P_1' = \frac{P_1}{\eta_N} \times 2 = \frac{105}{0.935} \times 2 = 224.6\text{kW}$$

2) The power loss caused by the surplus flow is

$$\Delta P = P_1' \times \frac{30}{130} = 224.6 \times \frac{30}{130} = 51.83\text{kW}$$

$$\Delta E' = \frac{\Delta P}{P_1'} = \frac{51.83}{224.6} = 23\%$$

That is to say, the power loss caused by back-flow is 51.83 kW, accounting for 23% of the total electricity consumption, from which it can be seen that the energy wastage is very serious. If installing a frequency converter in one of the motors, the surplus flow can be eliminated by reducing the speed, which would save this part of energy. But the frequency converter itself consumes energy, so the total power-saving rate is less than 23%. In this case, the combined pole changing speed control energy-saving methods also can be applied.

### 5.6.2 Energy-saving Principles of Combined Pole Changing Speed Control

In the process of water injection shown in Figure 5.16, the main trunk outlet pressure should meet the range of the given pressure, which ranges between 4 and 5 MPa; therefore, it is unnecessary to adjust the main trunk outlet pressure smoothly. First, the plunger pump characteristics will be introduced before discussing the combined pole changing speed control.

The output volume of the plunger pump at rated speed is a fixed value, independent of the pump outlet pressure and its own characteristics, determined by the pipeline system. Those are two independent parameters, and the output volume of the plunger pump is proportional to the speed.

$$Q = K \times n \tag{5.31}$$

where $Q$, the output volume of plunger pump ($m^3/h$);
$n$, the shaft speed of plunger pump (r/min), that is, the motor speed;
$K$, the factor.

When the plunger pump speed is constant, the shaft power is proportional to the outlet pressure

$$P_e = K \frac{p_Z \times n}{3.6\eta} \tag{5.32}$$

where $P_e$, the shaft power of plunger pump (kW);
$p_Z$, the outlet pressure of plunger pump (MPa);
$\eta$, the efficiency of plunger pump.

From Equation (5.31), it can be seen that the output volume of the plunger pump is proportional to the motor speed; changing the speed can change the output volume of the plunger pump, so when adjusting the output volume to be consistent with the demand (injection allocation rate), it would neither change the efficiency of the pump, nor affect the outlet pressure, which creates the condition of using combined pole changing speed control for the plunger pump. The following takes pumping station shown in Figure 5.16 as an example to introduce the combined pole changing speed control methods of motors. The method of motor transformation is as follows:

1) The pump motor 3 is changed into 6/10 poles, and the 6-pole power is 75 kW, 10-pole power is 45 kW;
2) The pump motor 4 is changed into 8/12 poles, and the 8-pole power is 55 kW, 12-pole power is 37 kW.

Each motor is equipped with a control cabinet, whose insides are equipped with detection and control devices, air switch, AC contactor, and so forth. There is a pressure sensor in the main line outlet, which forms a closed loop control system. It regulates speed by combined pole changing speed control method according to the outlet pressure of the main line. Based on the output volume requirement, the two double-speed variable pole motors totally have eight kinds of combinations. The power of motors and output volume of plunger pumps corresponding to those combinations are as follows:

Combination One: putting into 12 poles, the motor power is 37 kW, and the pump output volume is 21.7 $m^3$/h;
Combination Two: putting into 10 poles, the motor power is 45 kW, and the pump output volume is 26.0 $m^3$/h;
Combination Three: putting into 8 poles, the motor power is 55 kW, and the pump output volume is 32.5 $m^3$/h;
Combination Four: putting into 6 poles, the motor power is 75 kW, and the pump output volume is 43.3 $m^3$/h;
Combination Five: putting into 10 and 12 poles, the sum of the motor power is 82 kW, and the sum of the pump output volume is 47.7 $m^3$/h;
Combination Six: putting into 10 and 8 poles, the sum of the motor power is 100 kW, and the sum of the pump output volume is 58.5 $m^3$/h;
Combination Seven: putting into 6 and 12 poles, the sum of the motor power is 112 kW, and the sum of the pump output volume is 65 $m^3$/h.
Combination Eight: putting into 6 and 8 poles, the sum of the motor power is 130 kW, and the sum of the pump output volume is 75.8 $m^3$/h.

Obviously, the combinations of 6 and 8 poles or 8 and 12 poles are not included in those combinations.

Through the combinations above, the output volume regulation of the pumping station is smoother, which can meet the requirements of the actual working conditions in most cases. In the above pump stations, the main line pressure of the outlet is about 4–5 MPa, and the water injection volume is 100 $m^3$/h. In this condition, the pump 2 can be opened, and then the pump motor 4 of 8-pole winding is put into operation, the total output volume is 97.5 $m^3$/h, the total motor power is 165 kW. In such conditions, the total output volume has little deviation, which has little effect on working conditions. In another example, the requirement of the main line pressure of the outlet is 4–5 MPa and the water injection volume is 75 $m^3$/h; so pump motor 3 of 6-pole winding is put into operation, then the pump motor 4 of 8-pole winding is in operation, so the total output volume is 75.8 $m^3$/h, the total motor power is 130 kW.

The output volume of plunger pump is proportional to the speed; if using variable frequency speed control, a smooth adjustment of the output volume can be obtained. If using the combined pole changing speed control, the adjustment of output volume would be stepwise. The output volume and speed curve of two speed-control methods are shown in Figure 5.17, Curve 1 in the figure is the result of variable frequency speed control, and Curve 2 is the result of pole changing speed control. It can be seen clearly that the eight steps of the combined variable frequency speed control have a better

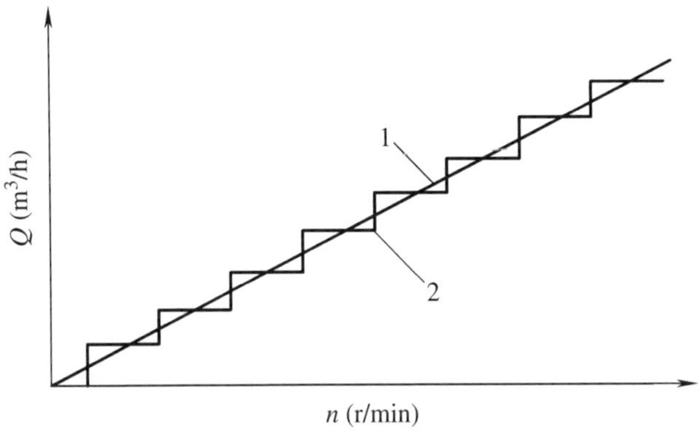


**Figure 5.17** Output volume and speed curve. 1, Frequency control; 2, pole changing speed regulation.

approximation to the variable frequency speed control curve, which can achieve a good energy-saving effect.

If changing two motors into three-speed pole changing motors, the output volume adjustment will be smoother; however, motor performance will be affected. According to the design principles of the motor, the slots coordination of dual-speed motors' stator and rotor are relatively easy to be taken into account, while three-speed motor is difficult, and the performance of one certain number of poles is greatly affected. In addition, the transformation cost of a three-speed motor is relatively high, and the performance price ratio is not high. Therefore, it is usual to regulate speed by the combined pole changing two-speed motors. Of course, the approximation effect of the output volume and speed by using the combined pole changing speed control method of the three-speed motor would be better.

### 5.6.3 Energy-saving Examples of Combined Pole Changing Speed Control

A water pumping station has three high-pressure three-plunger pumps of 51 mm in diameter, the output volume of each injection pump is 17.3 $m^3$/h in theory, and the actual output volume is 12.5 $m^3$/h. Normally, there are two pumps in operation, and one for spare. The power of motor that matched with the three injection pumps is 110 kW, and has 4 poles. The rated speed of the motor is 1470 r/min. Outlet pressure of injection pump is 19–21 MPa. Normally the output volume of opening one pump is not enough while excess of putting two pumps into operation, and the pressure is too high, reaching to 25 MPa; therefore, it needs to rely on back-flow to control the injection allocation rate and the outlet pressure of mainline. Also the quantity of back-flow is too large, causing a lot of energy wastage.

The energy-saving transformations of pumping station are as follows: converting two of those motors into dual-speed pole changing motors, one is 6/10 poles, the corresponding power is 75 kW/45 kW, the corresponding output volume of 6 poles is 8.33 $m^3$/h, and the corresponding output volume of 10 poles is 5.0 $m^3$/h; the other is 8/12 poles, the corresponding power is 55 kW/37 kW, the corresponding output volume of 8 poles is 6.25 $m^3$/h, and the corresponding output volume of 12 poles is 4.17 $m^3$/h. The two motors are both dual-winding pole changing motors. Each motor is equipped with a control cabinet, which can cast windings of different number poles according to the requirement. First, determine which one of the combinations in accordance with the mainline outlet pressure and the injection allocation rate, then operate the switching control of each winding. In practical problems, the injection pressure does not change frequently, and a condition can be stable for at least 1 or 2 months, so manual control can meet the need of actual condition. The total investment of the above transformation is 48,000 yuan.

The injection allocation rate of original pump station is 450 cubic meters daily ($m^3/d$). Before the transformation of motor, it needs to open two pumps, pump outlet output volume is 600 $m^3/d$ and back-flow is 150 $m^3/d$. The electricity consumption of pump station is 169,412 kWh per month. After the transformation, a motor of 110 kW (4 pole) is put in operation, and then a motor of 55 kW (8 pole) is put in. Pump outlet pressure is 21 MPa, output volume is 450 $m^3/d$, and back-flow is not needed. The electricity consumption of pump station is 127,750 kWh per month. The saving electricity is 41,662 kWh monthly, energy-saving rate is 24.6%. The electrical price of the pumping station is 0.60 yuan/kWh, saving 24,997 yuan a month, and the full investment can be recovered in 2 months.

If the pump station uses variable frequency speed control, at least 10 million or more is needed to be invested in one time. The variable frequency speed control motor should be chosen, that is, a YVF2 series converter asynchronous motor as shown in Table 2.16. Otherwise, when the motor speed is too low, the motor heats seriously, easy to be burned down. Thus, the transformation cost of variable frequency speed control will be higher.

### 5.6.4 Comparison of Combined Pole Changing Speed Control and Variable Frequency Speed Control

The combined pole changing speed control and the variable frequency speed control have their own advantages, the application is determined according to the specific circumstances.

1) The smoothness of the flow regulation of combined pole changing speed control is worse than variable frequency speed control. Under the working conditions that the requirements are not very strict, the combined pole changing speed control method can be chosen.
2) The costs of the combined pole changing speed control are lower, which is about 30% of the variable frequency speed control costs.
3) The reliability of the combined pole changing speed control is higher. The combined pole changing speed control mainly uses the AC contactor to accomplish switching control of the windings, and the control system is simple and reliable.
4) The efficiency of the combined pole changing speed control is higher than variable frequency speed control. In the cases of addressing the same surplus flow, the efficiency of the combined pole changing speed control is higher than the variable frequency speed control, since the frequency converter itself has 5–10% loss.
5) There is no harmonic pollution problem of the combined pole changing speed control, which is superior to a variable frequency speed control. Harmonics would

increase the supplementary losses of the motor, making the motor's temperature go higher.

Currently, harmonic pollution has become a major nuisance to the grid. Therefore, when using the pole changing or the combined pole changing, speed control could meet (or almost meet) the actual working conditions, and the variable frequency speed control should not be used as possible.

The combined pole changing speed control method is not only suitable for the energy-saving technological transformation of the old pumping station, but also for the new pumping station design. For new pumping station, as long as changing two motors into pole changing speed control motors, the investment would be reduced significantly. Combined pole changing speed control can not only be applied to an oilfield, but it can also be used for other fluid delivery systems. The application of combined pole changing speed control methods in other fields needs further study.

## 5.7 Discussion Topics in the Chapter

1) What is the energy-saving principle of speed matching?
2) In the fluid delivery system, when the demand is less than supply, if the motor is in constant speed operation, it will produce excess energy for the motor, rather than lower load-carrying.
3) What are the main speed control methods of three-phase asynchronous motors?
4) Why can the energy conservation of the variable frequency speed control system be calculated based on the pump operating point?
5) Draw the control block diagram of water supply system, explain the control theory, and explain the principles of constant pressure water supply system.
6) In a variable frequency speed control system, how to select a frequency converter.
7) What are the advantages and disadvantages of variable frequency speed control?
8) What are the advantages and disadvantages of pole changing speed control? Under what circumstances can the pole changing speed control be applied?
9) What is the energy-saving principle of the combined pole changing speed control? Try to illustrate which circumstance can also use the combined pole changing speed control.

# 6

# Energy-saving Principle and Method of the Mechanical Properties Fit

Energy conservation of the motor is a systematic project; it needs to consider not only the energy saving of the motor itself but also their mutual cooperation among the motor driving systems consisting of its driven loads. In this chapter, the energy-saving principles and methods of the reasonable adaptation between the mechanical properties and the load characteristics would be further discussed, after the discussion on the power adaptation and speed adaptation between the motor and the load in Chapter 5. Therefore, this chapter will take the beam-pumping unit currently widely used in oil fields as an example to study the energy-saving principle of the mechanical properties adaptation. First, the working principle, load characteristics, and mechanical properties of the beam-pumping unit are introduced; second, the mechanical properties of ultra-high slip motors, as well as the characteristics of winding motors with self-adaption of the rotor resistance are introduced; last, the energy-saving principles of the mechanical properties fit between the ultrahigh slip motor and pump loads are introduced, as well as the energy-saving examples of mechanical properties fit.

## 6.1 Load Characteristics of A Beam-Pumping Unit

Petroleum and petrochemical industries are the key of energy-saving electrical industries. The "energy-saving project of motor system" in the "ten major energy-saving projects" issued by State 8 ministries makes the petroleum and petrochemical industries key reform areas. Unlike the oil fields of the Middle East where most of the oil wells are flowing, the majority of Chinese oil fields are stripper wells. In most of the oil fields, the crude oil is pumped from the underground oil reservoirs by pumping unit. Pumping process consumes a lot of power and exchanging oil by electricity is a common phenomenon in oil exploitation in China. The proportion of the electricity in oil exploitation cost is growing bigger, and the power consumed by the pumping unit has exceeded 40% of the total electricity consumption in oil production. At present, the pumping unit holdings is more than 10 million in China, total installed capacity of motors is about 3500 MW, and power consumption is over a hundred million kilowatt hour annually. In these pumping units, beam-pumping units are the most widely used, accounting for more than 70% of the total pumping equipment. Therefore, taking a beam-pumping unit as an example, the study on the energy-saving principle of the reasonable adaptation between the motors' mechanical properties

*Energy-saving Principles and Technologies for Induction Motors*, First Edition. Wenzhong Ma and Lianping Bai.

and the loads' mechanical properties is particularly important. The working principle of beam pumping units is introduced here.

### 6.1.1 Working Principle of the Beam-Pumping Unit

The beam pumping unit oil production system consists of three parts. First, the ground segment, that is, the beam-pumping unit, the basic structure of which is shown in Figure 6.1. It consists of a motor, gear box, four-bar linkage, crank, balance weight, wiggle stick, horsehead, hanging rope, rod, and other components. Second, the underground part, that is, the structures of the pump and the reciprocating pump are shown in Figure 6.2. It mainly consists of a pump cylinder, suction valve, pump valve plunger piston (piston), and discharge valve. It hangs on the end of the tubing in the casing. Third, the rod string, which consists of the steel bars of hundreds of meters to thousands of kilometers, connected with the pumping unit on the ground, in connection to the pump piston underground. The main oil extraction processes of the beam-pumping units is that: the motors are connected through belts and gear reducers, and drive crank doing nearly uniform circular motion, the crank drives the beam to swing up and down through the connecting rod, the horsehead ahead of beam moves up and down, the beam hanger on the horsehead is connected to the polished rod upper wellhead, and the polished rod inside wellhead is connected to the sucker rod. The up-and-down reciprocating movement of the pumping unit drive the up-and-down reciprocating motion of the piston, and lift oil from underground to ground.

The working principle of the pump is as follows. The upward movement of the sucker rod is called up-stroke (Figure 6.2a) and the downward movement of the sucker rod is called down-stroke (Figure 6.2b). During up-stroke, the sucker rod drives the piston up, closing the pump discharge valve due to the weight and oil pressure of liquids in the pipe, and the piston lifts the liquid upward. Meanwhile, the

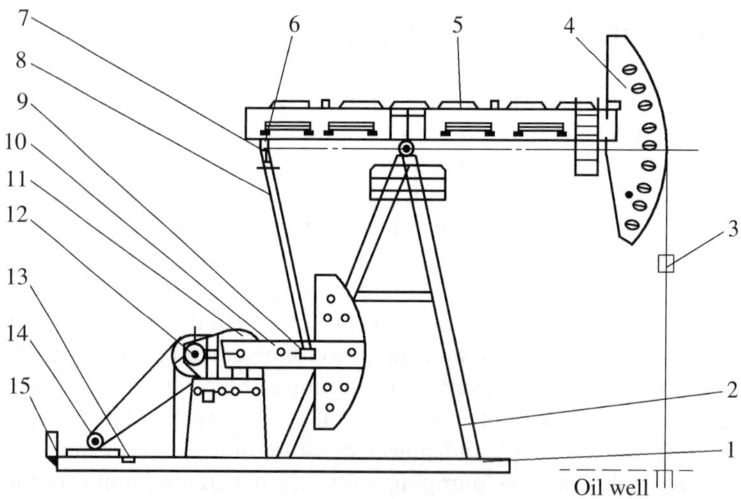


**Figure 6.1** The structure of a beam-pumping unit. 1, Pedestal; 2, support; 3, hanging rope; 4, horsehead; 5, wiggle stick; 6, beam bearing; 7, beam; 8, connecting rod; 9, crankpin device; 10, crank device; 11, gear box; 12, brake safety device; 13, brake device; 14, motor; 15, distribution box.

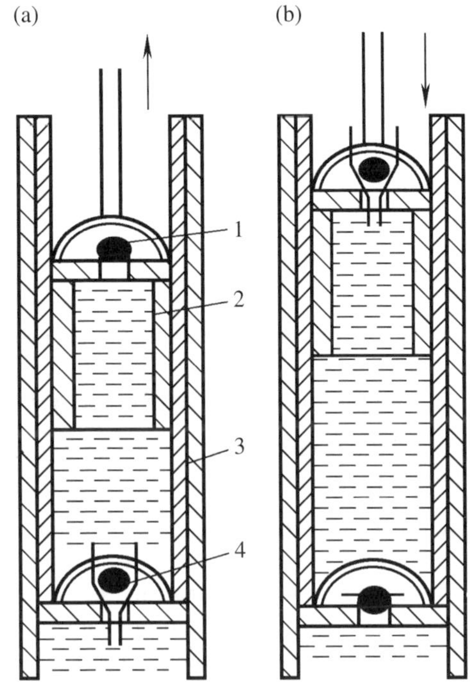


**Figure 6.2** The working principle diagram of reciprocating pump: (a) Up-stroke and (b) down-stroke. 1, Discharge valve; 2, plunger piston; 3, bushing; 4, suction valve.

pressure under the oil pump piston space is reduced, thus opening the suction valve under the pump, and the liquid is absorbed into the pump. During down-stroke, the sucker rod drives the piston downward, opening the pump discharge valve, and liquid flows into the upper portion of the piston pump. When the oil pump piston is driven up, this part of the liquid will be lifted again. Due to this continuously reciprocating motion of the pumping unit and piston pump, crude oil is poured on the ground. This is the oiling principle of the pumping unit, sucker rod, and the sucker rod–pump system.

The beam pumping unit converts the circular motion of the crank into a reciprocating linear motion of the sucker rod. During up-stroke, the pumping unit lifts the oil and the rod string, so its load is very heavy; during down-stroke, the gravity of the rod string will pull down the pumping unit. Balancing of measurements is required to overcome the big changes of the load in the process of the pumping unit operation. At present, the method commonly used in a beam pumping unit is adding balance weights on the crank to realize the balance of pump shown in Figure 6.1. Due to the restriction of the mechanical structure of the beam pumping unit, the load torque of the motor fluctuates seriously in a working cycle. Figure 6.3 is the torque diagram of a conventional beam pumping unit. As shown in the figure, the load torque has high peaks and deep valleys in a working cycle. The part below the horizontal axis indicates that the torque is negative; it means that the load drags running motor and the motor works in generating states. The large fluctuation of the torque characteristics would make a harsh requirement to the characteristics of its drive motor.

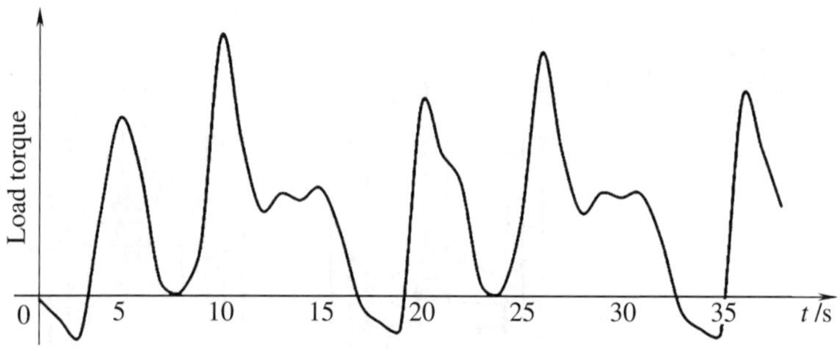


**Figure 6.3** The load torque curve of a beam pumping unit.

### 6.1.2 Requirements of Beam Pumping Unit to Drive a Motor

Through the front analysis, the beam pumping unit driving a motor must satisfy the following requirements:

1) In order to overcome the counterweight inertia, the motor should have a larger starting torque, making the pumping unit start smoothly.
2) Once the oil pumping machine is running, it just needs a small operating power, generally about one-third of the motor's rated power, or even lower. Therefore, the motor has acquired a wider efficiency curve and a power factor curve.
3) In a cycle, the average power of the motor is small, but the peak torque is large, so the motor must have enough instantaneous overload capacity.
4) In a cycle, the motor will have a period of time in the generating states, and generating electricity would be fed back to grid.
5) The motor speed should be reduced during the transformation period of the up-stroke to down-stroke of a pumping unit, in order to slow down the reversing impact on the sucker rod, reducing the elastic deformation of the sucker rod and prolonging the actual schedule of the pump piston.

Requirements (1) and (2) are contradictory; choosing a larger power motor can meet the requirements of starting power, but having a very low load rate in operation, the motor would be in "light load drive" status, resulting in a lower efficiency and power factor. This is the main reason of the low efficiency of the beam pumping unit system.

## 6.2 Energy-saving Principle of Mechanical Properties Fit

From the perspective of the slip, three-phase asynchronous motors can be divided into three kinds. The slip of an ordinary asynchronous motor is 2–5%; a high slip motor is 5–10%; and an ultra-high slip motor is 10–30%. At present, the driven motor that fits the mechanical properties of the beam pumping unit is mainly an ultra-high slip motor; the mechanical properties of ultra-high slip asynchronous motor are described in the following text.

### 6.2.1 Characteristics of an Ultra-High Slip Motor

Ultra-high slip motors, just as the name implies, are motors with a high slip and soft mechanical properties (Figure 6.4). The rotor cage of the ultra-high slip motor is made

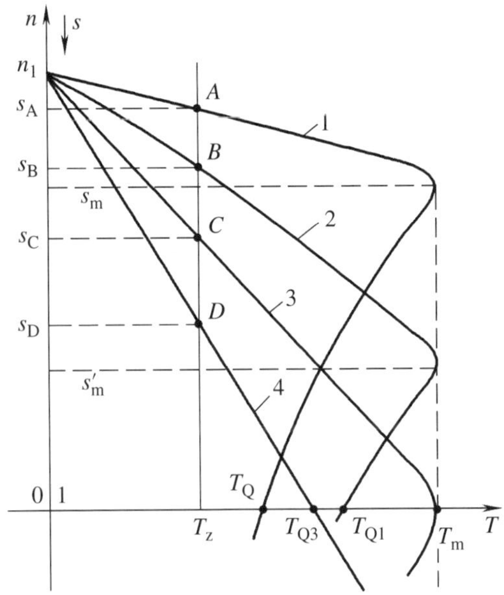


**Figure 6.4** The mechanical characteristic curve of the ultra-high slip motor.

of copper, aluminum, magnesium, and other alloy materials, increasing the resistance of the rotor, thus increasing the slip of the motor. The motor speed drops in a heavy load, and the torque increases; the speed increases in a light load, and the torque decreases, the speed range becomes larger, starting current is low, and starting torque is large.

Curve 1 is the mechanical property curve of a normal motor; Curves 2, 3, and 4 represent the mechanical characteristics of different rotor resistance values of a high slip motor which increase in turn. From the figure, it can be found that the mechanical properties of an ultra-high slip motor are very similar to the ordinary wound-rotor asynchronous motor of series string resistance in the rotor. For the operation of a wound-rotor asynchronous motor with series resistance in the rotor, if driving a constant torque load, as the load $T_Z$ in Figure 6.4, the motor will run in several different working points of A, B, C, and D; when the rotor resistance value varies, the slips increase in turn. When the slip increases, the power factor of the motor is constant but the efficiency falls. The following is the analysis of the change of the motor power factor, efficiency, losses, and starting torque with the change of slip.

#### 6.2.1.1 Analysis of Power Factor

Power factor is related to motor parameters, so the equivalent circuit of the asynchronous motor in Figure 2.10 needs to be observed again. As can be seen from the figure, when the slip changes, the stator branch and the excitation branch are not affected; it only affects the equivalent of rotor branch. The power factor angle of the rotor side can be obtained from the equivalent circuit.

$$\varphi_2 = \arctan\left(\frac{x'_{2\sigma}}{r'_2/s}\right) \tag{6.1}$$

Therefore, to analyze the power factor changes, the key is to analyze the change of $r'_2/s$.

Take point A and point B in Figure 6.4 as an example to analyze the change of power factor. Point A is the low slip operating point, and point B is the high slip operating point. When the slip varies, the load torque is constant, by the torque equation

$$T = \frac{m_1 p U_1^2 \frac{r_2'}{s}}{2\pi f_1 \left[\left(r_1 + \frac{r_2'}{s}\right)^2 + \left(x_{1\sigma} + x_{2\sigma}'\right)^2\right]} \tag{6.2}$$

It can be seen that only when $r_2/s_A = r_2'/s_B$, $T_A = T_B$ is guaranteed.

Therefore, when the motor is running with a constant torque load, $r_2'/s$ is constant when the slip changes, so the power factor angle of the rotor side $\varphi_2$ and $\cos\varphi_2$ remain unchanged; thus, the power factor obtained from the grid input also remains the same.

#### 6.2.1.2 Efficiency Analysis

The definition of motor efficiency is

$$\eta = \frac{P_2}{P_1} = \frac{T_2\Omega}{3U_1 I_1 \cos\varphi_1} = \frac{T_2 \times \frac{2\pi n}{60}}{3U_1 I_1 \cos\varphi_1} \tag{6.3}$$

Figure 6.1 shows that the torque of a constant torque load is constant, the speed $n$ reduces when using a ultra-high slip motor, and the output power $P_2$ decreases. On the other hand, because the constant $r_2/s_A = r_2'/s_B$, the parameters of the motor equivalent circuit are the same as in Figure 2.10: the input voltage $U_1$ is constant; $I_2$, $I_1$ are constant, and so the input power is constant. Therefore, the efficiency $\eta$ decreases.

#### 6.2.1.3 Loss Analysis

As $I_2$ is constant, the rotor resistance increases, $P_{Cu2} = 3I_2^2 r_2'$, the copper loss increases. In addition, since $P_{Cu2} = sP_{em}$, when the slip increases, the copper loss of the rotor (slip power) also increases accordingly.

#### 6.2.1.4 Analysis of Starting Performance

The starting torque formula of the motor is

$$T_{st} = \frac{m_1 p U_1^2 r_2'}{2\pi f_1 \left[\left(r_1 + r_2'\right)^2 + \left(x_{1\sigma} + x_{2\sigma}'\right)^2\right]} \tag{6.4}$$

As can be seen from formula (6.4), with the increase of rotor resistance, the starting torque is also increased.

The starting current formula is

$$I_{1st} \approx I_{2st}' = \frac{U_1}{\sqrt{\left(r_1 + r_2'\right)^2 + \left(x_{1\sigma} + x_{2\sigma}'\right)^2}} \tag{6.5}$$

From the above formula, the equivalent resistance of the rotor increases, and the starting current decreases.

Through the analysis above, when the ultra-high slip motor is applied in a constant torque load, the efficiency drops, there is no energy-saving effect, on the contrary, costing much more energy. Only the combination of an ultra-high slip motor and special loads (beam pumping unit) can obtain a better energy efficiency effect.

## 6.2.2 Energy-saving Principle of the Adaptation of Mechanical Properties

In case of an ultra-high slip motor used in a driving beam pumping unit, the energy-saving principle has the following two aspects.

### 6.2.2.1 With High Starting Torque, Lowering Power Level, Improving the Load Factor

The locked-rotor torque ratio of an ultra-high slip motor is 2.8–3.5 and the locked-rotor current ratio is 3.0–5.0; therefore, the ultra-high slip motor has excellent starting performance. When an ultra-high slip motor is used in a driving pumping unit load, on the one hand, the ordinary motor of larger capacity can be replaced by a smaller one, realizing reasonable power fit. It can solve the problem of "light load drive" well, and improve the operating efficiency of the motor, which has a good energy-saving effect. On the other hand, the decrease of the starting current can reduce the transformer capacity. For example, the transformers of oil well measurement stands supply power to the motors of six oil wells originally; after using the ultra-high slip motor, it can meet the electricity of seven to eight oil wells' motors. When the number of oil wells is constant, using small capacity transformers can not only save the equipment investment, but also reduce the power loss in the transformers.

### 6.2.2.2 Soft Features of Ultra-High Slip Motor Can Improve Coordination and Efficiency of the System

Mechanical properties of an ultra-high slip motor are soft features. In the process of dragging the pumping unit, when the load torque increases, the motor speed will drop; when the load torque reduces, the motor speed rises. As can be seen from Figure 6.3, the change of the pumping unit load torque is very complex, and the relatively flat period is less than a third in a cycle. For such a complex change of load, why need soft characteristics drive? This problem also should be analyzed from the power transmission mechanism of a pumping unit. The power transmission of a pumping unit is relying on the long rod. The sucker rod is connected by steel bars of a few meters long one by one. For a steel rod it can be regarded as a rigid rod; for thousands of meters of steel bar, it is no longer a rigid rod, rather an elastic rod, or rather a long spring. After such an elastic rod, the motion law of piston pump is no longer the ground movement law of the polished rod of the beam pumping unit. Along with the sucker rod, tubing friction, and viscous resistance of crude oil, the movement of the pump piston is very complicated. The main cause of the sucker rod elastic deformation is acceleration of the polished rod. Pumping unit from up-stroke to down-stroke or from down-stroke to up-stroke is called reversing. Pumping unit translates a circular motion into a reciprocating linear motion; in the processes of commutation and intermediate movement, there is always a change of acceleration, and therefore the sucker rod elastic deformation is inevitable. An effective measure is that driven by the ultra-super high slip motor of soft characteristics, slowing down the change of the acceleration, reducing the elastic deformation of the sucker rod, thus extending the actual schedule of subsurface pump plunger, improving the fill factor of pump, and increasing the producing fluid volume.

Using an ultra-high slip motor driving a beam pumping unit can not only achieve good energy-saving effect, but can also prolong the life of the pumping system. First of all, a high slip motor can significantly reduce the maximum torque of a pumping unit reducer, and extend the life of the reducer. Second, it would reduce the maximum stress and the change of stress, decrease the sucker rod break and rod off accidents, prolong the service life of the sucker rod, extend the service life of the equipment, and improve production efficiency.

Through reasonable adaptation of mechanical characteristics between the motor and load, it can improve the system fit, improve the efficiency of system, and increase production in the case that motor energy consumption remains the same. This is the principle of the mechanical properties fit.

### 6.2.3 Applications and Standards of Ultra-High Slip Motor

The United States has been studying high slip motors as early as the 1960s; an ultra-high slip motor was used in a beam pumping unit in 1970s. In 1980s, the effect of adopting a high slip motor to drive beam pumping units was affirmed internationally and the ultra-high slip motor obtained a lot of promotion, which has resulted in a very good system energy-saving effect. Most of the domestic applications of high slip motors are imported technologies. One is American technology and the other is Russian technology, both technologies having their own characteristics. American technology focuses on solving the problem of the ultra-high slip motor's dissipation: the material used in the rotor conductor is aluminum alloy, the strong heat dissipation ability of aluminum alloy can be used to solve the problem of motor temperature rising. Russian technology focuses on improving motor winding insulation level, improving the temperature resistant performance can achieve a higher slip of the motor. In the early 1980s, China started to introduce high slip motor technology, the Changed petroleum machinery factory formally batched production CJT-3 type and CJT-4 type pumping unit energy-saving drive apparatus in 1985. The device (YCCH) was composed of a high slip motor and its control box (CK). Through the promotion and application in the north China, Liaohe, and other oil fields, it was fully proven that using this device to drive a pumping unit can achieve two direct economic benefits of power saving and increases in output.

At present, the production of ultra-high slip motors has achieved a seriation and standardization, the specifications of the products is more completely expanded. Now the rated capacity of producing a motor is small to 3 kW, big to 55 kW. The ultra-high slip motor-rated voltages include 380, 660, 1140 V, and so on. The frequency is power frequency of 50 Hz, meeting the different needs of users.

The American Petroleum Institute (API) as early as in 1993 made the related standards of an ultra-high slip motor: API Spec 11 l6-1993 standard of motors for Beam Pumping Unit (Electric Motor Prime Mover for Beam Pumping Unit Service), and in 2008 was revised (API Spec 11 l6-2008 second edition). In 2005, China also issued the law of the People's Republic of China oil and gas industry standards SY/T 6636–2005 "standard of beam pumping unit USES." The formulation, distribution, and implementation of these standards laid a solid foundation for specification development of a high slip motor.

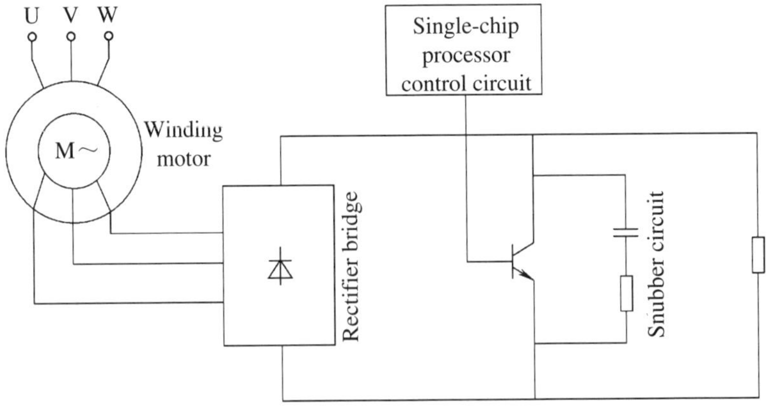


**Figure 6.5** The winding motor with an adaptive rotor resistance.

### 6.2.4 Applications of a Winding Motor

An ultra-high slip motor has achieved the energy-saving property of the adaptation system of load mechanical properties, but it also has its shortcomings. Excessive heat of motor rotor is the biggest flaw. Many manufacturers use aluminum to make motor shell, plus temperature control switch to control the temperature of the motor, to prevent motor burning because of overheat. In addition, rotor resistance of an ultra-high slip motor is fixed, and the mechanical properties cannot be adjusted. In term of these problems, the author studied the winding motor with adaptive rotor resistance. The system structure is shown in Figure 6.5; its working principle is that a three-phase rectifier bridge is connected in the output termination of an ordinary wound-rotor slip ring; followed by the rectifier bridge is a high-power resistance, a power electronic switching device is in parallel with the resistance (IGBT). The single-chip processor control system automatically detects the working state of the pumping unit. By calculation, the output control signal controls the IGBT on–off with pulse width modulation (PWM) mode; constituting the modulation chopper resistor, the rotor resistance can be automatically adjusted according to the working states of the pumping unit, thus changing the mechanical properties of the motor winding can achieve a good fit with the characteristics of pumping load, and improve system efficiency. At the same time the motor rotor resistance heating is on the outside of the motor, so the motor will not be burned down because of overheat of the rotor. This approach has achieved good results in the practical application of energy saving.

## 6.3 Energy-saving Instances of Mechanical Properties Fit

A type 10 beam pumping station, pump setting depth is 1500 m, load starting torque is large, red beats is 4-stroke per minute, liquid production capacity is 32 $m^3$ daily. The model of an original three-phase asynchronous driven motor is Y250M-6, rated power is 37 kW, and its average load rate of field test is 25%. The efficiencies of motor and pumping system are not high. Now after technological upgrading, three-phase asynchronous motors are replaced by ultra-high slip motors. The starting torque of an ultra-high slip motor is much higher than an ordinary motor; hence, the choice of an ultra-high slip motor of power 22 kW can meet the load requirements. Check the original three-phase asynchronous motor (Y250M-6) technical parameters $\eta_N = 90.8\%$, $\cos\varphi_N = 0.86$, $Q_0 = 16.65\text{kvar}$, and $\Delta P_0 = 1.504\text{kW}$. Look up 22 kW ultra-high slip motor (YCJT-200L2-6) technical parameters. $\eta_N = 84\%$, $\cos\varphi_N = 0.837$, $Q_0 = 9.46\text{kvar}$, and $\Delta P_0 = 1.25\text{kW}$. After using ultra-high slip motors, the load mainly remains the same, output power of the motor is constant, the motor load factor increases up to 42%, and the daily yield increased up to 35 $m^3$. The following analysis is about the comprehensive efficiency of the motor energy-saving effect before and after transformation, as well as the overall energy-saving effect of the pumping system. Since the secondary side of the power substation has central reactive power compensation, the reactive economic equivalent coefficient is selected to $K_Q = 0.05$.

### 6.3.1 Power Factor and Comprehensive Efficiency of Motor Before Transformation

The output active power of the motor is $P_2 = \beta P_N = 9.25\text{kW}$

The total active loss in operation is $\Delta P = \Delta P_0 + \beta^2 \left[ \left( \frac{1}{\eta_N} - 1 \right) P_N - \Delta P_0 \right] = 1.644\text{kW}$

The input active power of the motor is $P_1 = \beta P_N + \Delta P = 10.894\text{kW}$

The operating efficiency of the motor is $\eta = \frac{P_2}{P_1} \times 100\% = 84.9\%$

The rated reactive power of the motor is $Q_N = \frac{P_N}{\eta_N} \tan\varphi_N = \frac{37}{0.908} \times 0.5934 = 24.179\text{kvar}$

The input reactive power of the motor is $Q_1 = Q_0 + \beta^2 [Q_N - Q_0] = 17.121\text{kvar}$

The power factor of the motor is $\cos\varphi = \frac{P_1}{\sqrt{3}U_1 I_1} = \frac{P_1}{\sqrt{P_1^2 + Q_1^2}} = 0.537$

The comprehensive power loss of the motor is $\Delta P_c = \Delta P + K_Q Q_1 = 2.5\text{kW}$

The comprehensive efficiency of the motor is $\eta_c = \frac{\beta P_N}{\beta P_N + \Delta P_c} \times 100\% = 78.72\%$

### 6.3.2 The Power Factor and Comprehensive Efficiency of Switching 22 kW Ultra-High Slip Motor

After using 22 kW ultra-high slip motor, load is constant, the load rate increases to 42%.

The output active power of the motor is $P_2 = \beta P_N = 9.25\text{kW}$

The total active loss in operation is $\Delta P = \Delta P_0 + \beta^2 \left[ \left( \frac{1}{\eta_N} - 1 \right) P_N - \Delta P_0 \right] = 1.769\text{kW}$

The input active power of the motor is $P_1 = \beta P_N + \Delta P = 11.019\text{kW}$

The operating efficiency of the motor is $\eta = \frac{P_2}{P_1} \times 100\% = 83.95\%$

The rated reactive power of the motor is $Q_N = \frac{P_N}{\eta_N} \tan \varphi_N = \frac{22}{0.84} \times 0.6538 = 17.123\text{kvar}$

The input reactive power of the motor is $Q_1 = Q_0 + \beta^2 [Q_N - Q_0] = 10.812\text{kvar}$

The power factor of the motor is $\cos\varphi = \frac{P_1}{\sqrt{3}U_1 I_1} = \frac{P_1}{\sqrt{P_1^2 + Q_1^{\,2}}} = 0.714$

The comprehensive power loss of the motor is $\Delta P_c = \Delta P + K_Q Q_1 = 2.31\text{kW}$

The comprehensive efficiency of the motor is $\eta_c = \frac{\beta P_N}{\beta P_N + \Delta P_c} \times 100\% = 80.02\%$

### 6.3.3 Energy-saving Effect of Motor

The input active power and reactive power before transformation are $P_1 = 10.894\text{kW}, Q_1 = 17.121\text{kvar}$
The input active power and reactive power after transformation are $P_1' = 11.019\text{kW}, Q_1' = 10.812\text{kvar}$
Then the active power saving is $\Delta P_p = 10.894 - 11.019 = -0.125\text{kW}$
The reactive power saving is $\Delta Q = 17.121 - 10.812 = 6.309\text{kvar}$
The comprehensive energy saving is $\Delta P_c = \Delta P_p + K_Q \Delta Q = -0.125 + 0.05 \times 6.309 = 0.19\text{kW}$
The comprehensive energy-saving rate is

$$\Delta E_c = \frac{\Delta P_c}{P_{1q} + K_Q \times Q_{1q}} \times 100\% = \frac{0.19}{10.894 + 0.05 \times 17.121} \times 100\% = 1.62\%$$

From the above calculations, it can be seen that the active power electricity saving is negative after adopting an ultra-high slip motor. The motor itself not only saves energy, but also increases the energy consumption. The reactive power is about 6.309 kW; considering the reactive economic equivalent, the comprehensive energy-saving power after the combination of both motors is only 0.19 kW, the comprehensive energy-saving rate is 1.62%, so the energy-saving effect is not obvious.

### 6.3.4 Overall Energy-saving Effect of the Pumping Unit System

As the mechanical properties of an ultra-high slip motor are soft features, it can improve the coordination of the pumping unit system, and improve the fill factor of the defueling pump, thus raising the fluid production. According to the above calculation, the motor input power changes little before and after the replacement of motor, and the efficiency increase of production can be calculated directly.

$$\Delta E' = \frac{W_2 - W_1}{W_1} \times 100\% \tag{6.6}$$

where $\Delta E'$, the energy-saving effect of the pumping unit system;
$W_2$, the production after the technical transformation;
$W_1$, the production before the technical transformation.

Substituting production before and after the technical transformation into the above equation, we obtain the following:

$$\Delta E' = \frac{35-32}{32} \times 100\% = 9.4\%$$

The total energy-saving effect is

$$\sum E = \Delta E' + \Delta E_c = 9.4\% + 1.62\% = 11.02\%$$

Although energy saving of a motor itself is not obvious, the soft features of an ultra-high slip motor plays a very important role in driving the beam pumping unit system. Many similar instances in the oil field production, usually the efficiency of the pumping unit system, would increase by about 10%; hence, the energy-saving effect of some ultra-high slip motor would reach up to 15%.

To sum up, energy saving of motor should not only consider the energy saving of motor itself, but also the interaction between the motor and the load. If mechanical properties and load properties of motor are fitted reasonably, it can improve the coordination and efficiency of the system, and implement energy saving from the aspect of the system. Since the load mechanical properties vary widely, some loads require the coordination of a motor's hardened characteristics, whereas some loads require the coordination of a motor's soft characteristics, choosing which features of motor should be determined according to the actual situation of the dragging system.

Sometimes, selecting a motor drive system may not be to save energy, but to meet the requirements of the actual working condition. For example, an electric drill motor needs soft characteristics. The speed decreases when the load suddenly increases, torque increases, and the drill pipe would not twisted off; an excavator motor also needs soft features; the mining speed decreases when faced with stone or roots to avoid damaging the earth-moving machinery.

## 6.4 Discussion Topics in the Chapter

1) There are many energy-saving instances of mechanical properties adaptation in the actual production, try to illustrate it.
2) Is the mechanical properties of an ultra-high slip motor softer and better in the beam pumping unit driving system?
3) Why does the active power consumption of the motor increase itself after adopting an ultra-high slip motor drive pumping unit?
4) Adding the adaptive rotor resistance on the slip ring of a wound-rotor motor, can it achieve a better energy-saving effect than a high slip motor? Try to analyze the reasons.
5) Using an ultra-high slip motor to derive a beam pumping unit, in addition to saving energy, what are the other benefits?

# 7

# The Energy-saving Principle of Induction Motor Reactive Power Compensation

It is well known that the *t* induction motor can be regarded as inductive load, which not only consumes active power but also absorbs reactive power from the grid during operation. While the reactive power produces additional losses in the transmission line, the reactive power compensation should be considered in the induction motor's energy saving. This chapter will provide an introduction to the energy-saving principle of induction motor reactive power compensation, the capacity selection of the reactive power compensation capacitor, and the static and dynamic reactive power compensation of an induction motor. Meanwhile, a hybrid compensation method for the induction motor is proposed in the chapter. In addition, the three-phase induction motors discussed in the chapter are all under low voltage, usually about 380 V.

## 7.1 Energy-saving Principle of Induction Motor Reactive Power Compensation

Induction motors are the largest energy consumption facilities in the grid, which consume about 60–70% of the total reactive power. Many motors are running without load or with light load, so their load rate and power factor are both very low. The lower the load rate, the lower the power factor, and lower power factor means higher reactive power and higher loss of the line. So reactive power compensation for induction motor is an important way to improve the power factor and save electricity.

### 7.1.1 Reactive Power of Induction Motor

The simplified equivalent circuit of induction motor is shown in Figure 7.1. In the figure, $X_m$ refers to the excitation reactance, $X\sigma$ refers to the leakage reactance, and $R_s$ refers to the equivalent resistance for a particular load. The reactive power absorbed by the motor can be calculated by the equivalent circuit. Thus

$$Q = Q_m + Q_\sigma = \frac{U^2}{X_m} + I_\sigma^2 X_\sigma \tag{7.1}$$

*Energy-saving Principles and Technologies for Induction Motors*, First Edition. Wenzhong Ma and Lianping Bai.

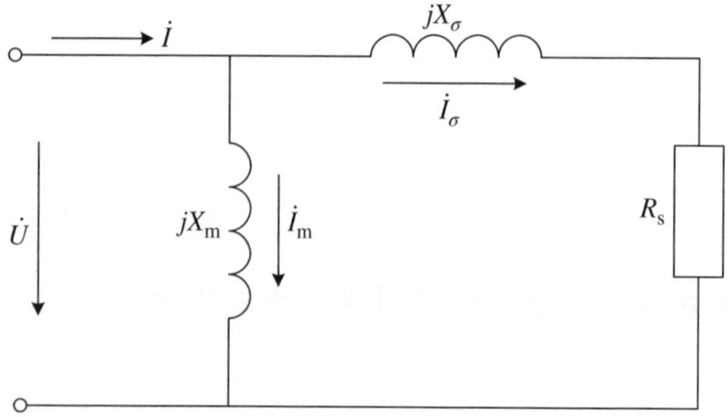


**Figure 7.1** Simplified equivalent circuit of induction motor.

where $Q_m$, the excitation reactive power (kvar), or excitation power for short;
$Q_\sigma$, the reactive power (kvar) absorbed by leakage reactance;
$X_m$, the excitation reactance;
$X_\sigma$, the leakage reactance.

The reactive power of the induction motor includes the excitation power and the reactive power absorbed by leakage reactance. The excitation power is proportional to the square of the voltage. Due to the effects of magnetic saturation, when the voltage increases, the excitation reactance declines. As a result, the excitation power changes with the voltage but is slightly above the quadratic curve. $Q_\sigma$ is the reactive power of the leakage reactance $X_\sigma$. The excitation power is an essential requirement for normal operation of the three-phase induction motor, which inevitably diverts the reactive power of the leakage reactance, so the power supply must provide reactive power during induction motor running.

### 7.1.2 Energy-saving Principle of Induction Motor Reactive Power Compensation

Take a single phase winding as an illustration. The equivalent circuit shown in Figure 7.1 can be further simplified to an equivalent circuit that consists of a resistor and an inductor in series. The transmission line can also be equivalent to a resistance–inductance series circuit, as shown in Figure 7.2. In order to compensate the reactive power required by the motor, a capacitor is applied in parallel connection at both ends of the motor windings, as shown in Figure 7.2. $\dot{I}'_L$ in the figure can be resolved into active component $I_p$ and reactive component $I_q$. Figure 7.3 shows a phasor diagram of induction motor with shunt capacitor. The horizontal coordinate of Figure 7.3 is active current $I_p$, and the reactive current, reactive power, and power factor before using shunt capacitors are $\cos\varphi_1$, $I_q$, and $Q_1$.

After using the shunt capacitance, the active current $I_p$ and the active power $P$ remain unchanged. The reactive power of capacitance is $Q_c$. The capacitor current is $I_c$, whose direction is in contrast to the reactive current component $I_q$. As shown in Figure 7.3, the reactive current after compensation is $I_{q2} = I_q - I_c$, the reactive power after compensation is $Q_2 = Q_1 - Q_c$, and the power factor after compensation becomes $\cos\varphi_2$. After using shunt capacitance, the power factor and reactive power are

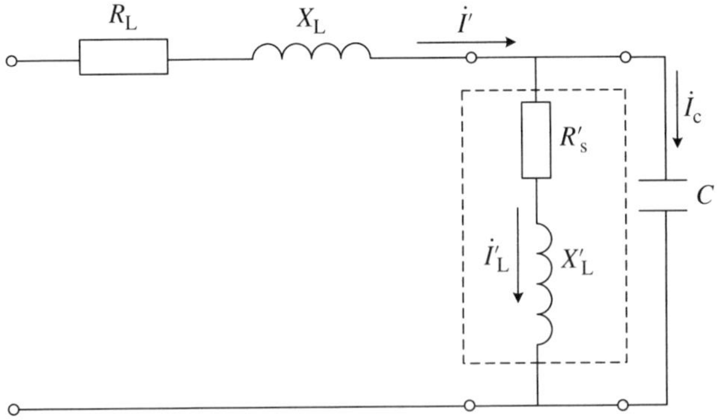


**Figure 7.2** Reactive compensation circuit of induction motor.

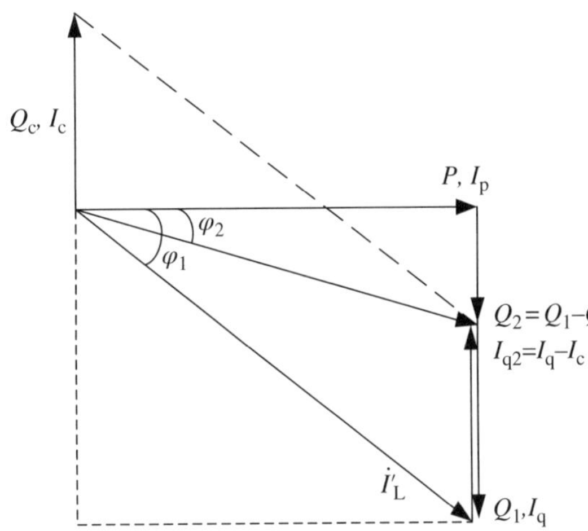


**Figure 7.3** Reactive power compensation phasor diagram of induction motor.

$$\cos\varphi_2 = \frac{I_p}{\sqrt{I_p^2 + \left(I_q - I_c\right)^2}} \tag{7.2}$$

$$Q_2 = P\left(\tan\varphi_1 - \tan\varphi_2\right) \tag{7.3}$$

Note: Capacitor compensation changes the current outside motor's external terminals, but the motor's internal current is not changed.

From the above analysis, it can be figured out that the current in the transmission line after paralleling capacitance is reduced, so does the transmission line's resistance loss, that is $P_R = I^2R$ becomes smaller, which is the energy-saving principle of reactive power compensation.

**Example 7.1** There is a 30 kW/8-pole three-phase induction motor, connected to the power transformer through a 50 m cable. The transformer's capacity is large enough, and its rated output voltage is 390 V. The resistance and reactance of cable are 0.2 and

0.15 Ω, respectively. The measured parameters of the motor in operation are as follows: the current is 46.6 A, power factor is 0.744, the efficiency is 0.92, reactive power is 20.5 kvar, and load rate is $\beta = 70\%$. Please calculate the appropriate value of the shunt capacitors, and calculate the line losses and line voltage drop before and after the reactive power compensation.

**Solution**

Referring to Figure 7.2, before using the shunt capacitance, the line active power loss is

$$\Delta P = I^2 R_L \times 3 = 46.6^2 \times 0.2 \times 3 = 1302.9$$

and the line voltage drop can be calculated as

$$\Delta U = \frac{PR + QX}{U_N} \tag{7.4}$$

where $P$, the motor active power;
$Q$, the motor reactive power;
$R$, the line resistance;
$X$, the line choke;
$U_N$, the transformer rated output voltage.

The active power of the motor is

$$P = \beta \frac{P_N}{\eta_N} = 30 \times \frac{0.7}{0.92} = 22.826\text{kW}$$

Substituting various parameters into (7.4), the line voltage drop equals to

$$\Delta U = \frac{PR + QX}{U_N} = \frac{22{,}826 \times 0.2 + 20{,}500 \times 0.15}{390} = 19.6\text{V}$$

Since the motor's reactive power in actual test is 20.5 kvar, according to capacitor serial standards, it is appropriate to select the 20 kvar parallel capacitor. Thus, the power factor after paralleling capacitor is

$$\cos\varphi = \frac{P}{\sqrt{P^2 + (Q_1 - Q_2)^2}} = \frac{22{,}826}{\sqrt{22{,}826^2 + (20{,}500 - 20{,}000)^2}} \approx 1$$

The line current after paralleling capacitor is

$$I' = \frac{P}{\sqrt{3}U\cos\varphi} = \frac{22{,}826}{1.732 \times 371.4 \times 1} = 35.5\text{A}$$

The active loss of line after paralleling capacitor is

$$\Delta P' = (I')^2 R_L \times 3 = 35.5^2 \times 0.2 \times 3 = 756.2\text{W}$$

The line voltage drop after paralleling capacitor is

$$\Delta U' = \frac{PR + QX}{U_N} = \frac{22{,}826 \times 0.2 + 500 \times 0.15}{390} = 11.9\text{V}$$

It can be seen that the active line loss reduces 546.7 W and the line voltage drops reduce 7.7 V after paralleling capacitor.

### 7.1.3 Role of Induction Motor Reactive Power Compensation

Reactive power compensation for induction motor will bring good economic benefits for enterprises.

1) *Good Energy-saving Effect*
   For the motors with larger capacity, operating with light load for a long period and with long power lines, using reactive power compensation on the spot can achieve better energy-saving effect.
2) *Lower Electricity Fee*
   In China, if the enterprises' power qualities are not up to the total power factor standard, they would be fined by utility companies according to the power factor regulating tariff measures. From the previous discussion, it is clear that local compensation for motors can improve power factor. The enterprises, whose power factors are high enough, would be rewarded.
3) *Higher Utilization Rate of a Transformer*
   The transformer's reactive current can be reduced through the induction motor reactive power compensation; thus the transformer's load will be lightened, and the supply capacity of the transformer will be correspondingly increased. If the company wants to expand production, the transformer' capacity is required to increase by 20%, and motor reactive power compensation in situ can completely meet this requirement under normal circumstances, so it does not need to purchase a new transformer and ancillary equipment, saving the investment costs.
4) *Higher Safety and Reliability*
   As adopting reactive power compensation, the currents of the electrical equipment and power lines reduce. Thus, the switch contacts failure and the supply line heat decline accordingly which improves the security and reliability.
5) *Better Starting Performance*
   After motor using reactive power compensation in situ, the line voltage drop is reduced. The terminal voltage after compensation becomes higher, and the electromagnetic torque will be increased; as a result, the starting performance of motors will be improved, especially for the motors with longer power lines.

### 7.1.4 Methods for Induction Motor Reactive Power Compensation

A three-phase induction motor is commonly compensated on the spot, namely, the compensation capacitor is connected in parallel to the motor's power ports. The method of reactive compensation adopted depends on the load operation mode. The properties of the load can be basically divided into the following three types:

1) General load, which is characterized by long working cycles, load changes slowly, power factor increases with the load's increase;
2) Rapid changing load, which is characterized by short working cycle, load changes fast, but the power factor still increases with the load's increase;

3) Special flicker load, which is characterized by a very short working cycle, the load is often working in the state of flicker, and the power factor is low. The most prominent example is the rolling mill and pumping machine; for this kind of load, conventional compensation method is generally difficult to work, new technologies, new theories, and power electronic devices should be adopted to solve the problem.

According to the above classification methods, we can conclude the reactive power compensation methods into the following three forms:

1) Static compensation, also known as delayed switching mode. This long time delay compensation, generally refers to the switching interval that is longer than 5 s, is mostly used in the situations where loads change slowly. This switching mode uses the AC contactor to turn on and off the capacitor.
2) Dynamic compensation, also known as instantaneous switching mode. Its delay time is very short. It can complete the switching action in tens of milliseconds. This switching mode mainly depends on power devices like thyristors combined with digital technology to control the capacitor switching. The dynamic compensation is mainly used in occasions of rapid changes of load or special flicker, like occasions using the welding machines, the pumping units in oil field, rolling mills, and so on. But the cost of a dynamic compensation device is much higher than the static compensation.
3) Hybrid compensation is a compensate mode which combines the static method and dynamic method. It uses the static way to compensate the main part of the total reactive power and then uses the dynamic way to compensate the fluctuation portion. The hybrid compensation has the advantage of low cost and better effect.

## 7.2 Capacity Selection for the Compensating Capacitor

If the load's fluctuation range is not wide, the induction motor usually uses the local static compensation. The key problem in this compensation is how to determine the capacity of the compensation capacitor. The reactive power calculation of the induction motor, the features of reactive power curve, and the determination of compensation capacitor capacity are introduced as follows.

### 7.2.1 The Calculation of Induction Motor's Reactive Power

The reactive power of the induction motor changes along with the load, but it is unchanged in the situation of rated load and no-load. Following is the analysis of its changing process.

1) Reactive Power at Rated Load

$$Q_N = \frac{P_N}{\eta_N} \tan \varphi_N \tag{7.5}$$

where $P_N$, the rated capacity of the motor (kW);
$\eta_N$, the motor efficiency at rated load;
$\tan \varphi_N$, the tangent of the rated power factor angle.

2) Reactive Power at No-Load $Q_0$

$$Q_0 = \sqrt{3U^2 I_0^2 \times 10^{-6} - P_0^2} \tag{7.6}$$

where $U$, the terminal voltage of the motor (V);
$I_0$, the no-load current of the motor (A);
$P_0$, the no-load loss power of the motor (kW).

Usually $Q_0$ accounts for 60–80% of the total reactive power at rated load. This is because the reactive power of the motor is mainly excitation power in the no-load situation.

3) The Reactive Power at Any Load $Q$

$$Q = Q_0 + \beta^2 (Q_N - Q_0) \tag{7.7}$$

where $\beta = P_2 / P_N$ represents motor load factor.
The $(Q_N - Q_0)$ can be defined as the reactive increment $\Delta Q$, that is,

$$\Delta Q = Q_N - Q_0 \tag{7.8}$$

Therefore,

$$Q = Q_0 + \beta^2 \Delta Q \tag{7.9}$$

It can be known from (7.9) that the reactive power of the motor at any load consists of two parts: one part is $Q_0$, it is a constant value that will not be affected by the load changing; another part is the variable part, it is proportional to the square of the load factor. Therefore, the variable part is only affected by changing of load. When the motor is selected, the unchanged part is constant, and the variable part is mainly affected by the load property and working mode. This feature should be taken into consideration sufficiently when implementing the reactive power compensation.

### 7.2.2 The Reactive Power Curve of Induction Motor

The three-phase induction motor's reactive power is related to not only the motor's rated power but also the number of poles.

Figure 7.4 shows the reactive power curves of 4-pole motors. Curves 1, 2, 3, 4, 5, 6, and 7 corresponds to the 15, 18.5, 22, 30, 37, 45, and 55 kW reactive power curves, respectively.

Figure 7.5 shows the reactive power curves of 6-pole motors. Curves 1, 2, 3, 4, 5, 6, and 7 corresponds to 15, 18.5, 22, 30, 37, 45, and 55 kW reactive power curves, respectively.

Figure 7.6 shows the reactive power curves of 8-pole motors. Curves 1, 2, 3, 4, 5, 6, and 7 correspond to 15, 18.5, 22, 30, 37, 45, and 55 kW reactive power curves, respectively.

From the above three figures, it can be seen that

1) for the motors with a small rated power, the reactive power curve is relatively flat, and the value of $Q_0/Q_N \approx 0.8$;
2) when the load rate is less than 60%, the reactive power curve is relatively flat;
3) when the load rate is greater than 70%, the reactive power curve change is relatively large;
4) for the motors with a large rated power, the reactive power curve change is relatively large, and the value of $Q_0/Q_N \approx 0.6$.

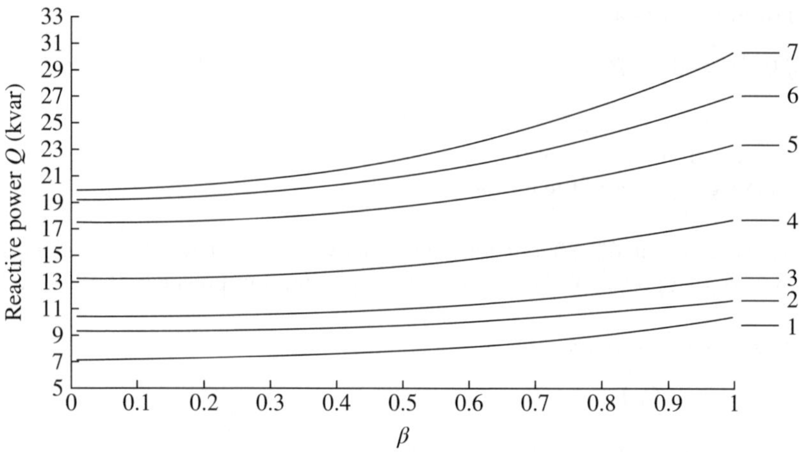


**Figure 7.4** The reactive power curve of 4-pole motors.

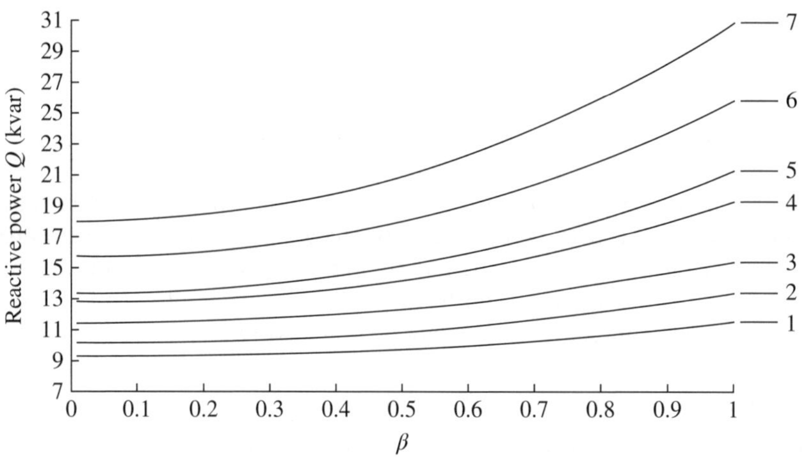


**Figure 7.5** The reactive power curve of 6-pole motors.

Curves 1, 2, and 3 in Figure 7.7 correspond to 4-, 6-, and 8-pole 30 kW motor reactive power curves, respectively. It can be seen from the figure that for the motor with the same rated power, due to the different number of poles, the variation of the reactive power is relatively large.

### 7.2.3 The Capacity Selection of the Induction Motor Compensation Capacitor

As can be seen from the reactive power curve, the reactive power of the induction motor changes with load rate, this change has a significant impact on the power factor. The capacity choice of reactive power compensation capacitor is not only affected by the motor's on-load reactive power $Q_0$ and the rated load reactive power $Q_N$, but also by the motor's load rate. In other words, the compensation capacitor's capacity should be chosen according to the range of the motor's load rate in its usual operation.

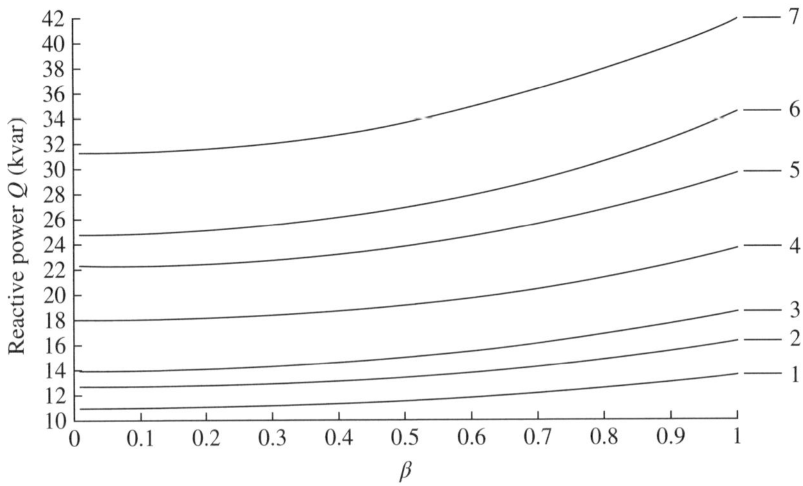


**Figure 7.6** The reactive power curve of 8-pole motors.

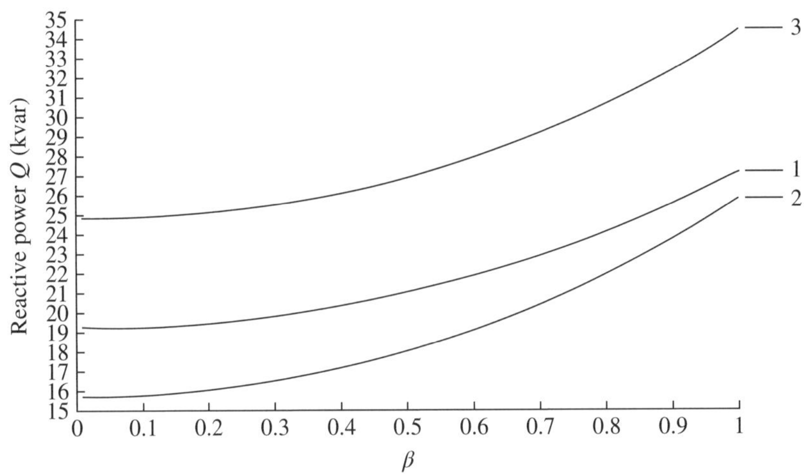


**Figure 7.7** 30 kW, 4-, 6-, 8-pole motors' reactive power curve.

**Example 7.2** A 22 kW/6-pole three-phase induction motor usually operates with the load factor between 60 and 80%. The measured value of unload reactive power of the motor $Q_0 = 11.40$ kvar, the rated reactive power $Q_N = 15.40$ kvar, the input power is 14.60 kW when the load rate is 60%, and the input power is 19.45 kW when the load rate is 80%. Try to select the capacity of the compensation capacitor.

**Solution**

The three-phase induction motor's reactive power at a usual load rate is

$$Q = Q_0 + \beta^2 (Q_N - Q_0)$$

Calculate the reactive power when the load rate $\beta = 0.6$ and $\beta = 0.8$, respectively.

$$Q_{0.6} = 11.4 + 0.6^2(15.4 - 11.4) = 12.84$$

$$Q_{0.8} = 11.4 + 0.8^2(15.4 - 11.4) = 13.96$$

In the vicinity of the value of the above calculations, according to the national standard, the values of low-voltage shunt capacitor's rated capacity are 10, 12, 14, and 15 kvar. First select the 12 kvar capacitor to do the calibration of the power factor. When the load rate is $\beta = 0.6$ and $\beta = 0.8$, the power factor of the motor is calculated, respectively, as follows:

$$\cos\varphi_{0.6} = \frac{P}{\sqrt{P^2 + \Delta Q^2}} = \frac{14.6}{\sqrt{14.6^2 + (12.84 - 12)^2}} = 0.998$$

$$\cos\varphi_{0.8} = \frac{P}{\sqrt{P^2 + \Delta Q^2}} = \frac{19.45}{\sqrt{19.45^2 + (13.96 - 12)^2}} = 0.994$$

In the range of 60–80% of the load rate, the power factor is high, which is close to 1. As a result, it is appropriate to choose the 12 kvar capacitor. But why do we choose the lower reactive power compensation capacitor in the above question instead of the higher? To explain the question, here take the 14 kvar capacitor to calculate the power factor again.

$$\cos\varphi'_{0.6} = \frac{P}{\sqrt{P^2 + \Delta Q^2}} = \frac{14.6}{\sqrt{14.6^2 + (12.84 - 14)^2}} = 0.996$$

$$\cos\varphi'_{0.8} = \frac{P}{\sqrt{P^2 + \Delta Q^2}} = \frac{19.45}{\sqrt{19.45^2 + (13.96 - 14)^2}} = 0.999$$

Within this load factor range, the compensation effect is also very good. However, there is overcompensation when the load rate is less than 80%. The overcompensation is bad for the motor: on the one hand it will increase the cost and on the other hand, it will increase the voltage of the circuit, thus increasing the system's instability. From the above analysis we can see that when the motor's load rate range from 60 to 80%, it is appropriate to choose 12 kvar capacitors.

Another way to choose the capacity of the compensating capacitor for the induction motor is the measure method. According to the motor actual operation, we can use the electricity comprehensive test instrument to measure the average value of the reactive power over a time. Then select a closed capacitor value in the down rank of the rated capacity level. This method is more practical and convenient.

### 7.2.4 Low-Voltage Shunt Capacitor

#### 7.2.4.1 Self-Healing Low-Voltage Shunt Capacitor

We usually use a low-voltage shunt capacitor of the power capacitors as the reactive power compensation capacitor. It is a whole film and dry-type low-voltage capacitors. It has the polypropylene film as medium, vacuum silver-plated or galvanized membrane as the electrode; after rolling and the vacuum treatment, it is set into a plastic box and sealed with a thermosetting resin, and is finally assembled in a steel box. The steel box is filled with nonflammable, inert, nontoxic natural mineral particles. The motor's local compensation device is usually installed in the vicinity of the motor; it should not only avoid the accidents and fires that are caused because of capacitor damage but also avoid the capacitor distributing toxic and harmful gases which is caused by the capacitor's aging. Therefore, after taking these treatment methods, such type of capacitor will not explode or burn, and does not produce harmful or toxic gases. Such capacitors have self-healing abilities, that is, when there is a local weakness in the capacitor film breakdown on the effect of the electric field, it will result in the discharge phenomenon, so that the surrounding metal sublimate and escape, and the capacitor immediately restores its function. Therefore, this type of capacitor is called a self-healing low-voltage shunt capacitor.

The characteristics of low-voltage self-healing shunt capacitor are low loss, low heat, low temperature rise; small size, light weight, good features; strong self-healing ability, long life; wide temperature range; relatively small capacity error; a certain overload ability; the cheaper price. Therefore, it is suitable for a three-phase induction motor local compensation.

The commonly used low-voltage self-healing shunt capacitor's model has the following meanings.

The model and its meaning

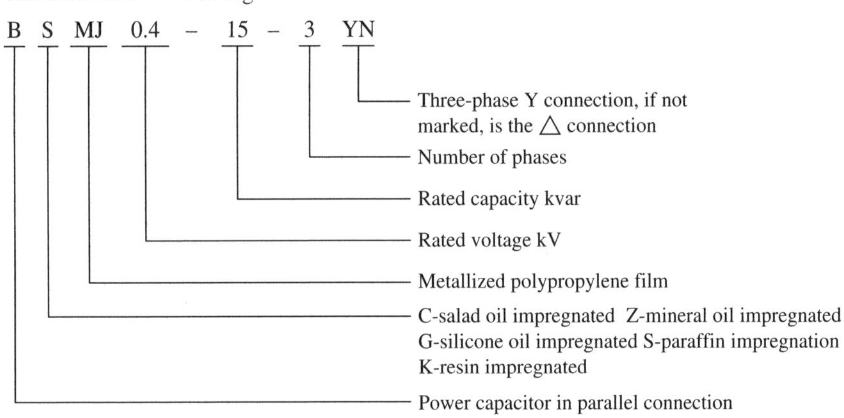


#### 7.2.4.2 Main Technical Indicators

1) The tangent of the loss angle: less than 1 W/kvar;
2) Capacitance tolerance: 0 to +10%;
3) Interelectrode withstanding voltage: $1.75U_N$ maintain 10 s;
4) Voltage between the electrode shells: 3000Vac maintain 10 s;
5) The maximum overvoltage: 110% of the rated voltage;
6) The maximum over-current: 130% of the rated current;
7) Insulation: 500 V DC applied between the electrode shells, for 1 min, the resistance greater than 100 MΩ;
8) The self-discharge characteristics: 1 min after disconnecting the power supply, the remaining voltage drops below 50 V.

#### 7.2.4.3 Environmental Conditions for the Operation

1) Temperature range: −25 °C–+50 °C;
2) Humidity range: less than 85% RH;
3) Altitude: no more than 2000 m;
4) Installation environment: dry, ventilated, noncorrosive gases and nondust, avoid the direct sunlight and erosion by rain and snow.

#### 7.2.4.4 Main Parameters of the National Standards

The selection of the shunt capacitors is mainly dependent on two parameters: one is the rated voltage, another is the rated capacity. The selection should meet the following national standards.

1) National standard GB12747-91 "The self-healing low-voltage shunt capacitor" provisions of the rated voltage's ratings are 0.23, 0.4, 0.525, 0.69, and 0.75 kV.
2) The national standard GB12747-91 "The self-healing low-voltage shunt capacitor" provisions of the rated capacity's ratings are 1, 1.6, 2, 2.5, 3, 4, 5, 8, 10, 12, 14, 15, 16, 20, 25, 30, 40, 50, 80, 100, 150, 200, and 334 kvar.

### 7.2.5 Research of Reactive Power Compensation for Induction Motor

In the experimental table, a test was made on a two-speed motor whose type is YD200L-4/6 and the power is 18.5 kW when 4 poles, and 15 kW when 6 poles. The test pole number is 4 and load of three phase induction motor is increasing from 0.18 kW to full. Each time increase the load by 3 kW, a total of eight power points. The input voltage, current, active power, and reactive power can be seen in Table 7.1.

As we can see the test figure from the table, no-load reactive power is about 7.31 kvar and load ratio $\beta = 13–73\%$ (namely output power varies from 2.44 to 3.55 kW), reactive power varies from 10.2 to 10.82 kvar. The change is relatively smooth. So we choose a compensation capacitor of 10 kvar. After shunting the capacitor, the test values of voltage, current, active power, reactive power, and power factor can be seen in Table 7.2. Here, the power factor is $\cos\varphi_2$.

From the test data in the table, we can see that only at no-load or load is light, the power factor is low, which equals to 0.207. When load ratio is no less than 13%, the power

**Table 7.1** The test data of the motor (18.5 kW/4 pole).

| Measuring parameter | $P_2$ | | | | | | | |
|---|---|---|---|---|---|---|---|---|
| | 0.18 | 2.44 | 5.34 | 8.07 | 10.85 | 13.55 | 16.22 | 18.45 |
| $U_1$ (V) | 396.9 | 396.3 | 394.0 | 391.1 | 390.4 | 387.8 | 386.2 | 385.0 |
| $I_1$ (A) | 10.62 | 15.54 | 17.32 | 20.20 | 23.60 | 27.60 | 31.91 | 35.74 |
| $P_1$ (kW) | 0.568 | 2.98 | 6.07 | 9.02 | 12.03 | 15.02 | 18.00 | 20.50 |
| $Q$ (kvar) | 7.31 | 10.20 | 10.22 | 10.32 | 10.45 | 10.82 | 11.60 | 12.11 |
| $\cos\varphi_1$ | 0.078 | 0.280 | 0.520 | 0.678 | 0.756 | 0.812 | 0.844 | 0.861 |

**Table 7.2** The test data of a motor with a compensation capacitor (18.5 kW/4 pole).

| Measuring parameter | $P_2$ | | | | | | | |
|---|---|---|---|---|---|---|---|---|
| | 0.18 | 2.44 | 5.34 | 8.07 | 10.85 | 13.55 | 16.22 | 18.45 |
| $U_1$ (V) | 396.9 | 396.3 | 394.0 | 391.1 | 390.4 | 387.8 | 386.2 | 385.0 |
| $I_1$ (A) | 3.96 | 4.37 | 8.96 | 13.42 | 17.96 | 22.60 | 27.20 | 31.10 |
| $P_1$ (kW) | 0.570 | 2.95 | 6.10 | 9.05 | 12.00 | 15.08 | 18.03 | 20.55 |
| $Q$ (kvar) | −2.73 | 0.23 | 0.27 | 0.32 | 0.48 | 0.85 | 1.23 | 2.17 |
| $\cos \varphi_2$ | 0.207 | 0.997 | 0.999 | 0.999 | 0.999 | 0.998 | 0.996 | 0.995 |

factor is near 1 after compensation. So the capacity of capacitor is very suitable. If 8 or 12 kvar capacitor was chosen, the effect will be worse, and the latter has a phenomenon of over compensation.

In fact, the above data comes from two sets of electric energy meter which were in parallel with the shunt capacitor and tested at the same time. There is torque and speed sensor on the experimental table, and the output power $P_2$ of the motor shaft can be tested directly. The minimum load of loading devices is 0.18 kW.

Reactive power curve of the motor before and after compensation can be seen in Figure 7.8. The reactive power is relatively small when load of two speed motor is less than 10%. Load ratio varies very little from 10 to 70% and it can be the reference to choose a capacitor.

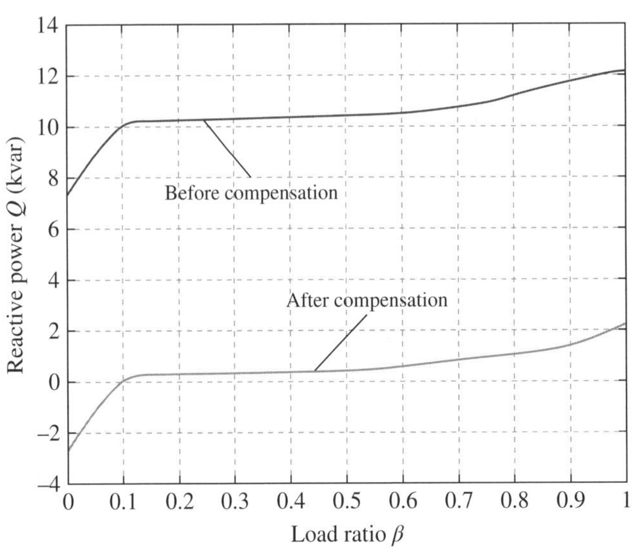


**Figure 7.8** 18.5 kW/4-pole motor's reactive power curve before and after compensation.

From the study above, selecting the capacitor capacity according to reactive power variation in motor's load ratio range will corresponds to reality and the compensation effect will be better.

### 7.2.6 Experiential Formula for Compensation Capacitor of Induction Motor

Over these years, the capacity selection of a reactive power compensation capacitor for induction motor has been extensively studied at home and abroad. Now introduce some experiential formula for the selection of capacity.

1) Ensure compensation capacitor by no-load current.

$$Q_c = \sqrt{3} U_N I_0 \tag{7.10}$$

where $U_N$, the rated voltage of the motor;
$I_0$, the no-load current of the motor.

As for $I_0$, which is difficult to find out, it can be calculated by the formula recommended by Sweden General Electric Company.

$$I_0 = 2I_N(1 - \cos\varphi_N) \tag{7.11}$$

where $\cos\varphi_N$, the rated power factor of the motor;
$I_N$, the rated current of the motor (A).

2) Compensation capacitor can be calculated according to 0.9 times of no-load current. In order to avoid self-excitation or overcompensation when the motor slows down, the compensation capacity can be calculated according to 0.9 times of no-load current, that is,

$$Q_c = 0.9\sqrt{3} U_N I_0 \tag{7.12}$$

3) When the motor power $P_N < 7.5\text{kW}$, the compensation capacity can be calculated as follows:

$$Q_c > 0.68 P_N \tag{7.13}$$

4) When the motor power $P_N \geq 7.5\text{kW}$, the local compensation capacity can be calculated as follows:

$$Q_c = 0.612 P_N \tag{7.14}$$

5) Local compensation capacitor of motor for machine tool is

$$Q_c = 2.7 P_N(\cos\varphi_2 - \cos\varphi_1) \tag{7.15}$$

where $\cos\varphi_1$, the power factor before compensation;
$\cos\varphi_2$, the power factor after compensation.

6) When the motor is not affected by fast reversal, reclosing, frequent starting, or other familiar operations which cause overvoltage or overtorque, the local compensation capacitor can be calculated from the following formula:

$$Q_c = \frac{1}{2} \times \frac{P_N}{\eta_N \cos\varphi_N} \tag{7.16}$$

7) Calculation of compensation capacitor by power factor standard is

$$Q_c = \frac{\beta P_N}{\eta_N}(\tan\varphi_1 - \tan\varphi_2) \tag{7.17}$$

where $\tan\varphi_1$, the tangent value of power factor angle before compensation; $\tan\varphi_2$, the tangent of the power factor angle after compensation.

8) The capacity of reactive power compensation for a water pumping motor is

$$\sqrt{3}U_N I_0 < Q_c \le \sqrt{3}U_N I_N \sin\varphi_N \tag{7.18}$$

In a word, in order to ensure that the capacity of local compensation is suitable, first the motor's parameters should be mastered entirely, including characteristic of the load, range of the load ratio, situation of running, safety location, and so on. Based on this, we can choose an appropriate capacitor.

## 7.3 Static Reactive Power Compensation of Induction Motor

The reactive power compensation capacitor of a three-phase induction motor is a constant value, not changing with the load, and that is what we usually called static compensation without power. In fact, selection of capacitor in the last section is aimed at the static compensation. The following will discuss the connection mode and relevant technology problems of static compensation.

### 7.3.1 Mode of Static Compensation

The static compensation capacitor is usually installed in the motor control cabinet, and the terminal of the motor is connected to the power output. The following will introduce five connection modes. In order to narrate conveniently, the three-phase lines of power source, motor, and the capacitor are all shown as one line, and the dotted line below stands for the control cabinet.

The first mode is connecting the capacitor to the terminal of the motor as shown in Figure 7.9. Figure 7.9a is a situation for motors that start/stop regularly, using a contactor to control, and can also be used in frequent start/stop situation. Figure 7.9b is for nonfrequent start/stop situations, using air switching to control. This mode is simple and only suitable for the motor whose power is less than 5 kW and load inertia is small.

Note: For the large inertia motor drive system, this connection mode is very dangerous. After the motor cuts off the power supply, the inertia will carry the motor to rotate. As the capacitor discharges to the motor for excitation, making the motor an induction generator and the electricity has no consumption circuit, causing the motor's voltage exceed several times of rated voltage, which is harmful for the insulation of motor and capacitor, and they are easy to burn out.

The second mode is that the capacitor is connected to the motor's terminal through a contactor, as shown in Figure 7.10. That is, the capacitor is in front of the contactor. Switching mode of the motor is the same as above. Figure 7.10a use the contactor as a

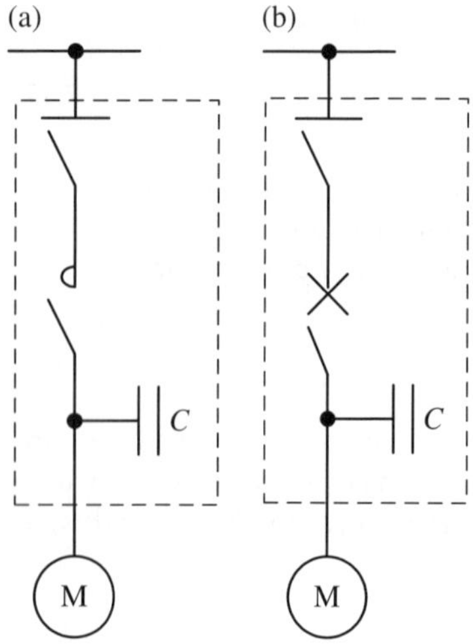


**Figure 7.9** Local compensation mode. (a) Contactor control circuit and (b) air switching control circuit.

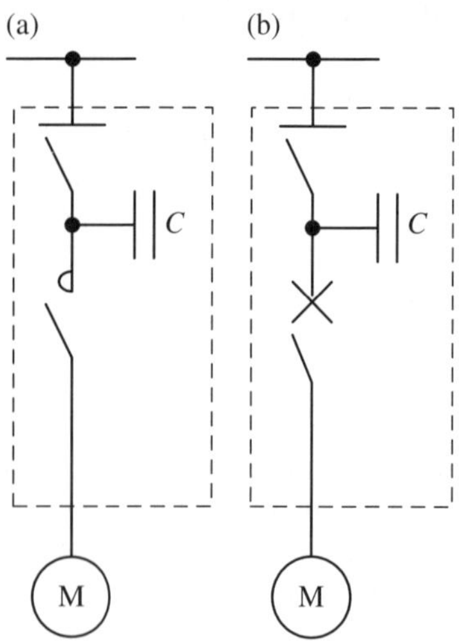


**Figure 7.10** Connection mode through contactor. (a) Contactor control circuit and (b) air switching control circuit.

switching controller, and Figure 7.10b use air switching. In this mode there is no capacitor discharging current working as the motor's excitation after the motor cutts off power supply, and this can prevent the motor running in self-excited generating state. However, after machine halt, the capacitor loses the discharge loop. After waiting for 3–5 min self-discharging, the self-healing low-voltage shunt capacitor will reach to a safety voltage.

The third mode is a protection mode which capacitor adopts fuse protection, as shown in Figure 7.11. In order to improve the reliability of the compensation device, a fuse protection device is often used. That is, adding fuse protection in front of capacitor. This connection mode is used for motors of special operations, such as reduced voltage starting or reversing operation. For these motors, if the capacity is small, you can also install capacitor in the starting cabinet or other controlling cabinets. This compensation mode is convenient and flexible, but the cost will increase and the problem of charged capacitor still exists.

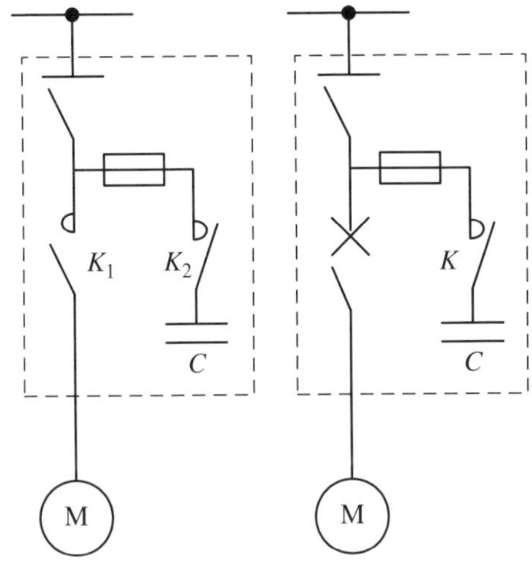


**Figure 7.11** A connection mode with fuse.

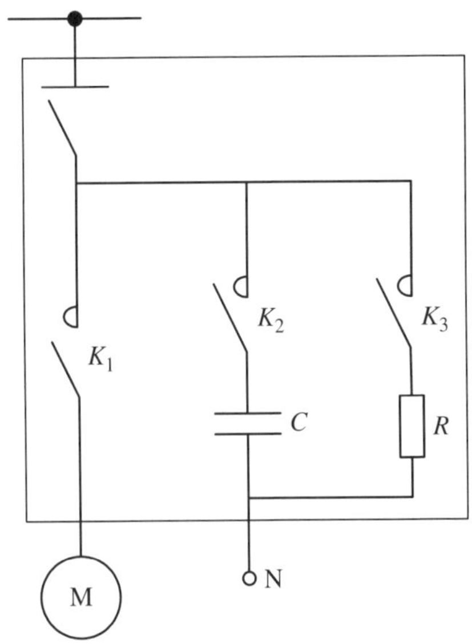


**Figure 7.12** A connection mode with discharging circuit.

The fourth mode is a compensation connection mode with discharge circuit, as shown in Figure 7.12. Input bleeder resistor after disconnecting the capacitor. Disconnect the resistor after charging. This mode fits for self-control compensation. It will add some cost for the excess of a contactor and a bleeder resistor, solving the problem of charging and self-exciting. For the condition that motor power is high, inertia loads is relatively large, and this compensation control mode can be adopted.

The fifth mode is a compensation connection mode with dash current suppression, as shown in Figure 7.13. The larger the motor's power (dozens of kilowatt) is, the larger the compensation capacitor will be. At the moment the capacitor puts in, dash current will be very large. In order to suppress dash current, you can put a reactor in series. When

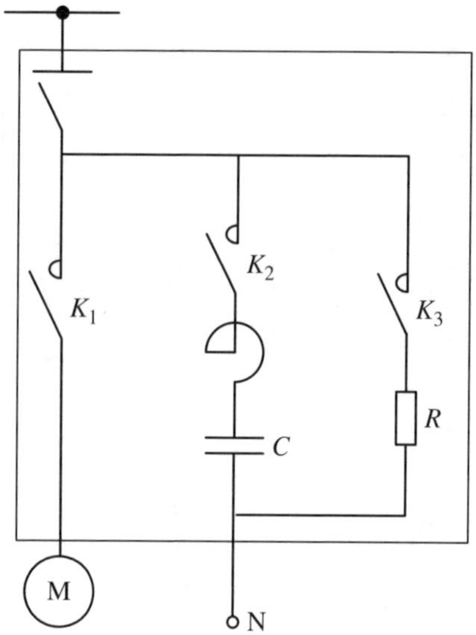


Figure 7.13 Connection mode of dash current suppression.

disconnecting, the bleeder resistor is still put in. Disconnect the resistor after discharging. This mode may cost more. However, the adding device is very necessary for the compensation of high power motors.

### 7.3.2 Caution for Static Compensation

#### 7.3.2.1 Prevent the Emerge of Self-Excitation

As previously mentioned, the motor adopting local capacitor compensation will continue rotating under the effect of inertia after power off. At the moment, the discharge current turns into exciting current; if the compensation capacity is enough, it will make the magnetic field of motor result in self-excitation and run in generation state. Motor and capacitor may be unable to endure the high voltage. In order to prevent the phenomenon of self-excitation, you should adopt the latter three connection modes above. Or you can reduce the capacity of compensation capacitor, and the formula of capacitor is as follows:

$$Q_c \leq 0.9\sqrt{3}U_N I_0 \tag{7.19}$$

#### 7.3.2.2 Overvoltage Protection

When a motor adopts reactive power compensation, the terminal voltage will rise, that is, the voltage of the installed capacitor will rise. The value can be calculated as follows:

$$\Delta U' = \frac{U_N Q_c}{S_s} \tag{7.20}$$

where $\Delta U'$, the increased value of motor's terminal voltage (V);
$Q_c$, the capacity of capacitor (kvar);
$S_s$, the short circuit capacity of the capacitor installation (kvar).

From formula (7.20), when the capacity of the compensation capacitor is large, the grid voltage will rise too much, even to the level which is not allowed. According to our national standard of parallel capacitor, overvoltage of power frequency should be less than 1.1 times the rated voltage. So when the capacity of a capacitor is relatively large, it must conform to the condition of $Q_c < 0.1S_s$.

#### 7.3.2.3 Prevent Overtime of Maintenance Voltage

The larger the capacity of the static compensation capacitor is, after the power source off, the longer the maintenance voltage time will keep. As for the motor which needs automatic switching, such as reversible motor, within the transition time from state of forward into reverse, the terminal voltage must decrease to 10% of rated peak voltage; otherwise it may cause self-excitation and terminal voltage will rise to the level of inadmissibility. When you use the motor windings as the discharge elements, the internal time $t$ should be no less than the value calculated below

$$t = 4.6\frac{L}{R} \tag{7.21}$$

where $L$, the inductance of motor winding (H);
$R$, the resistance of motor winding (Ω).

From formula (7.21), when motor windings are used for a discharge device of the static compensation of motor, the capacitor internal time $t$ is determined by the motor's structure parameters itself.

#### 7.3.2.4 Avoid the Resonance

The formula to avoid resonance is

$$n = \sqrt{\frac{S_s}{Q_c}} \tag{7.22}$$

where $S_s$, the short capacity of the capacitor installation site (MVA);
$Q_c$, the capacity of the compensation capacitor (Mvar).

If $n$ is integer, the capacitor will produce resonance in $n$ times harmonic wave. This situation can be avoided in means of adjusting the capacity of a capacitor.

#### 7.3.2.5 Prevent System Harmonic Influence

For the power supply circuit which has a harmonic source in the system, if the following conditions are met, you can believe that it has little effect on the capacitor. Otherwise, you must add reactor so that the harmonic effect will not harm the capacitor.

1) Capacity of thyristor device (kW) is not larger than 40% of the total load capacity (kVA) of substation.
2) The ratio of thyristor devices' kW value and the kilovolt-ampere value of short-circuit capacity of capacitor installation site is less than 0.05.
3) The device that generates harmonics, such as an electric arc furnace, is less than 25% of the transformer capacity. Or the ratio of the short capacity of capacitor installation set is less than 1/80.

Usually, the added reactor restricts the system's harmonic wave. When there is too much third harmonic in system, the electric reactor with 12% resistance rate is used to limit; when there is too much fifth harmonic in system, the electric reactor with 4.5% resistance rate is used to limit.

#### 7.3.2.6 Suppression of Capacitor Dash Current

The dash current of capacitor can be calculated by the following formula:

$$I_s = I_c\sqrt{\frac{2S_s}{Q_c}} \tag{7.23}$$

where $S_s$, the short circuit capacity of capacitor installation site (MVA);
$Q_c$, the capacity of the compensation capacitor (MVAR).

If $I_s$ is bigger than 100 times of $I_c$, you would better adopt measures to restrain the inrush current; if capacitor-dedicated AC contactors are used, you can use circuit breaker operating with resistance starting, or installing reactor.

When you design a static compensation scheme for a motor, the problems above should be noted and checked carefully. Besides, national standards "Parallel capacitor devices design specifications" (GB50227-2008) have many related regulations. When you design and check the compensation device of the motor, you have to contrast carefully to the related national standards. The detailed introduction is ignored.

### 7.3.3 Verification of the Static Compensation Capacitor

**Example 7.3** A factory used a Y series three-phase induction motor as an air compressor. The rated power is 55 kW, $U_N = 380$ V, rated efficiency is 92.5%, $\cos\varphi_N = 0.88$, $I_N = 104.9$ A, no-load reactive power $Q_0 = 18$ kvar, load rate varies from 80 to 100%, $S_N$ of power transformer equals to 500 kVA, the secondary rated voltage of the transformer is 400 V, impedance voltage $U_d\% = 4.5$, the compensation capacitor is installed in the contactor load side. Please try to calculate the capacity of the capacitor, and accounting each index.

**Solution**

1) Calculate Initial Capacity of the Compensation Capacitor
   According to formula (7.7), you can calculate the initial capacity of the compensation capacitor; first calculate

$$Q_N = \frac{P_N}{\eta_N}\tan\varphi_N = \frac{55}{0.925}\times 0.5397 = 32\text{kvar}$$

According to the principle of compensating low power factor, calculate reactive power when $\beta = 0.8$.

$$Q = Q_0 + \beta^2(Q_N - Q_0) = 18 + 0.8^2(32 - 18) = 26.96\text{kvar}$$

Refer to national standard GB12747-91 "self-healing type low-voltage parallel capacitor," it regulates the rated capacitor capacity, we can choose 25 kvar compensation capacitor as the initial capacity. Then calculate power factors when the rate load $\beta = 0.8$ and $\beta = 1$. The motor efficiency is higher when $\beta = 0.8$ than rated efficiency, so you can calculate motor efficiency by the rated efficiency.

$$P = \frac{\beta P_N}{\eta_N} = \frac{0.8 \times 55}{0.925} = 47.57\,\text{kW}$$

So

$$\cos\varphi_{0.8} = \frac{P}{\sqrt{P^2 + \Delta Q^2}} = \frac{47.57}{\sqrt{47.57^2 + 1.96^2}} = 0.999$$

$$\cos\varphi_{1.0} = \frac{P}{\sqrt{P^2 + \Delta Q^2}} = \frac{59.46}{\sqrt{59.46^2 + 7^2}} = 0.993$$

We can see that within the function scope of motor's rate load, the power factor is high.

2) Added Value of System Steady Voltage

$$\Delta U' = \frac{U_d\% \times U_N \times Q_c}{100 \times S_N} = \frac{4.5}{100} \times \frac{25}{500} \times 400 = 0.9\,\text{V}$$

The result will not affect the normal work of motor.

3) Resonant Condition

$$n = \sqrt{\frac{S_s}{Q_c}} = \sqrt{\frac{S_N / U_d\% \times 100}{Q_c}} = \sqrt{\frac{500 / 4.5 \times 100}{25}} = 21.08$$

Because $n$ is not integer, it will not produce resonance.

4) External Harmonic Influence

There is no great harmonic source, and the check is not necessary.

5) Dash Current

$$I_c = \frac{Q_c}{\sqrt{3}U_N} = \frac{25}{\sqrt{3} \times 380} = 38\,\text{A}$$

$$I_s = I_c\sqrt{\frac{2S_s}{Q_c}} = 38 \times \sqrt{\frac{2 \times 11{,}111}{25}} = 38 \times 29.81 = 1133\,\text{A}$$

The dash current is not more than 50 times of the rated current of a compensation capacitor, so you need not consider the current limiting measurements.

After the check, BSMJ0.4-25-3 is appropriate for the compensation capacitor. The connection mode of the static reactive compensation circuit is the discharge circuit, which is shown in Figure 7.13.

### 7.3.4 The Main Device Selection of the Compensation Device

#### 7.3.4.1 Selection of Discharge Resistance

Resistance is the main discharge device, and reactive compensation devices have a set of requirements for the discharge resistance. The choice of resistance should meet discharge time and safety requirements.

1) Basic Requirements of the Discharge Resistance of a Capacitor
   According to the relative national standards, there are three main requirements.
   - Capacitor should be an installed discharge device, so that the peak value of voltage in capacitor can fall to 50 V. As for the low-voltage capacitor, the longest time is 1 min.
   - When the capacitor may be disconnected and in a very short time be closed, with the capacitor on the voltage again, its terminal voltage cannot be higher than 10% of the rated voltage.
   - The loss of the discharge device should not exceed 1% of the rated capacity of capacitor.
2) Calculation of the Discharge Resistance
   The IEC published *Power Capacitor Installation and Operation Guide* recommended formula is as below

$$R = \frac{tU_N^2}{3.183Q_c \ln\left(\sqrt{2}U_N / U_r\right)} \tag{7.24}$$

where $t$, the discharge time (s);
$U_N$, the rated voltage of the capacitor (V);
$U_r$, the surplus voltage of the capacitor (V);
$Q_c$, the capacity of the capacitor (kvar).

#### 7.3.4.2 Selection of the Current Limiting Reactor

When the capacity of static on-site compensation capacitor is too large, you should consider using the reactor to prevent the inrush current. Hence, we will discuss how to select the reactor.

1) The Selection of Reactor's Capacity

$$Q_x = K_x Q_c \tag{7.25}$$

where $Q_x$, the capacity of the reactor;
$Q_c$, the capacity of the capacitor;
$K_x$, the ratio of the reactor. In the general system, when the third harmonic is more, $K_x$ is 12%, and fifth harmonic is 4.5%.

2) Selection of the Reactor's Rated Current
   For the three-faze reactor

$$I_{xN} = \frac{Q_x}{3U_{xN}} \tag{7.26}$$

where $I_{xN}$, the reactor's rated current;
$U_{xN}$, the reactor's rated voltage.

3) Reactance Loss
When the reactance ratio is 4.5%, and the loss rate is 0.013 kW/kVA. When the reactance ratio is 12%, and the loss rate is 0.015 kW/kVA. The noise level of 100 kVA and below of the oil immersed reactor is 46 dB.
4) Check of the Limited Inrush Current
The Inrush current calculation formula after series reactor is

$$I_{sm} = \sqrt{\frac{2}{K_x}} \times I_N \tag{7.27}$$

When $K_X = 4.5\%$, $I_{sm} = 6.7I_N$; when $K_X = 12\%$, $I_{sm} = 4.1I_N$.

This shows that the inrush current is limited as 6.7 times lower as the rated current when we use the reactor, and it is beneficial to the selection of the switch equipment. However, after the reactor, the end voltage of the capacitor will be higher than the original voltage. When the voltage is too high, it will prevent the capacitor operating with overvoltage.

#### 7.3.4.3 Contactor Selection

The AC contactor which is used for switching capacitor or air switch's rated current is

$$I_{cJ} = 1.5 \times 1.1 \times I_c = 1.65 I_c \tag{7.28}$$

where $I_{cJ}$, the air switch or contactor's rated current;
$I_c$, the capacitor's rated capacity.

In order to enlarge the ability to withstand the inrush current, $I_{cJ} = 2I_c$ is suggested. You can also use the special capacitor used for switching contactor.

#### 7.3.4.4 Fuse Selection

When using the fuse as the open circuit protection, you should limit the current peak value when operating 100 times lower as the rated current of capacitor.

## 7.4 Reactive Power Dynamic Compensation of the Induction Motor

Usually we use the AC contactor to switch the capacitance in the reactive power compensation of the induction motor. Sometimes, according to the power factor, it is switched gravely by the contactor. For the special load which is highly volatile in a cycle, we are unable to obtain satisfactory results by switching the contactor. For example, for the pumping unit used in an oil field, the motor load is always volatile and it works in the generator status in a cycle with a low average power factor. For this particular load, it is difficult to work with the static compensation on the spot or centralized compensation. Therefore, we must use the dynamic reactive power compensation method.

For dynamic reactive power compensation, there are many mature technologies, both at home and abroad. Besides, there are two main compensation methods. One is the dynamic reactive power compensation method which is based on TCR phase control, SVC compensation method for short (Static Var Compensator (SVC)); another dynamic reactive power compensation method is based on self-turn-off devices (such as IGBT), SVG compensation method for short (Static Var Generator (SVG)). Compared with the synchronous power adjustment motor, these two kinds of reactive power compensation are also known as static reactive power compensator whose devices are called static reactive power compensators.

### 7.4.1 Dynamic Compensation Based on TCR Phase Control

#### 7.4.1.1 The Circuit Theory of Transistor Phased-Control Dynamic Compensation

The principle of the phased transistor dynamic compensation is shown in Figure 7.14. Each group of antiparallel thyristors is connected in series with the capacitors, and then connects in Y type to make up the main circuit of the dynamic compensation. In order to restrain the inrush current, it is needed to string the reactor into the main circuit. The transistor phase-controlled dynamic compensation circuit also contains the synchronous phase detecting circuit, the microcomputer detection control circuit, and the sync pulse triggering circuit. We will, respectively, introduce the operating principle of each part below.

1) The Detection Circuit of Voltage and Current. In Figure 7.14, the detecting circuit consists of two parts: one part is synchronized phase detecting circuit (Figure 7.15), and another portion is the voltage and current detecting circuit. The phase detecting

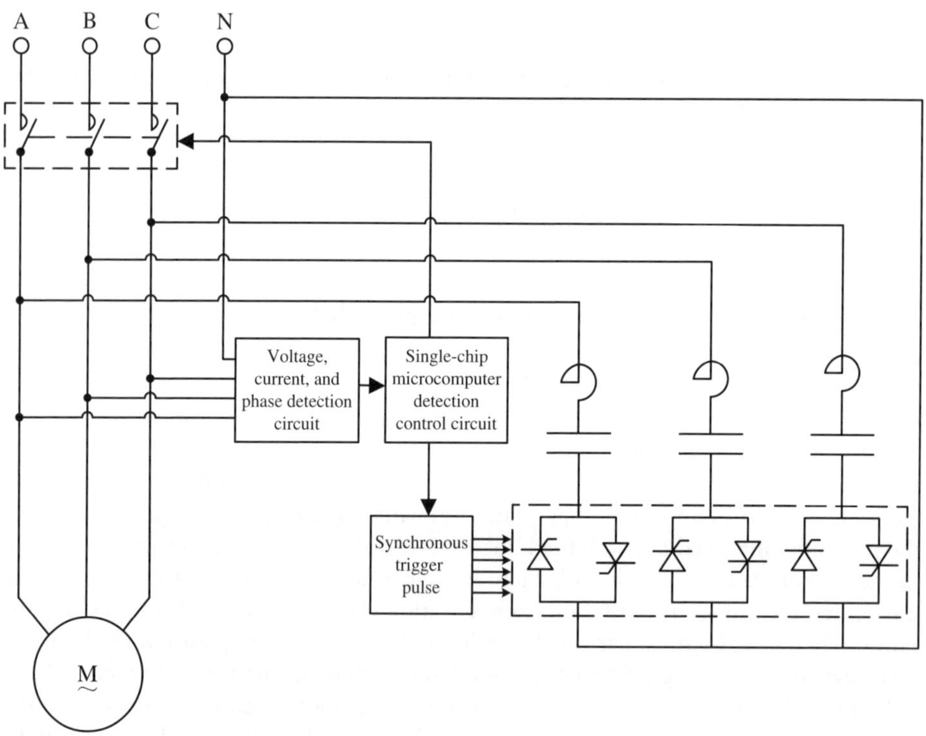


**Figure 7.14** The circuit theory block diagram of phased transistor dynamic compensation.

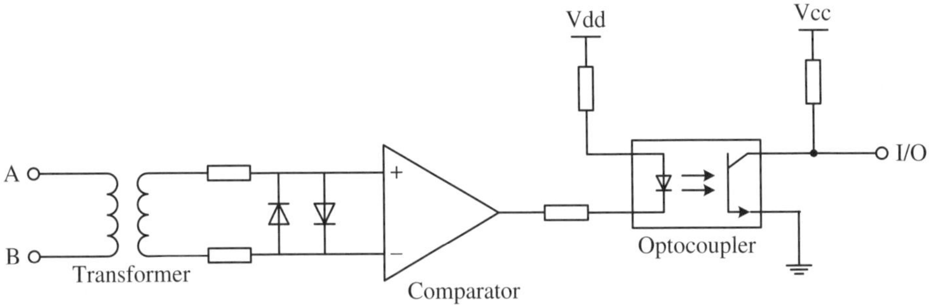


**Figure 7.15** The synchronous phase detecting circuit.

circuit detects the zero-crossing point of the U, V, and W line voltages, respectively, the figure only shows the U-phase. V phase and W phase are the same as the U-phase. The detected three line voltage's zero crossing a signal is sent to microcomputer through the optic-coupling isolation circuit. Voltage and current detecting circuit can be tested by a voltage transformer and a current transformer, the acquisition of the voltage and current signals are sent to the microcomputer after processing.

2) Microcomputer Detection and Control Circuit. In order to calculate the phase conveniently, the microcomputer with A/D acquisition function can be used. If it is a one-way A/D single chip, the voltage and current phase comparison circuit is also needed. According to the voltage, current, and the phase difference between the phase voltages and phase currents detected, a microcomputer calculates the reactive power of the motor in time $\Delta t$. That is,

$$\Delta Q = 3U_{\mathrm{p}} I_{\mathrm{p}} \sin \varphi s \tag{7.29}$$

where $U_{\mathrm{p}}$ and $I_{\mathrm{p}}$ are the motor phase voltage and phase current.

3) The Sync Pulse Triggering Circuit. The sync pulse triggering circuit is shown in Figure 7.16. The synchronization pulse triggering circuit consists of two main parts: one part is the synchronize pulse signal issued by the microcomputer and the other part is the pulse transformer. The pulse signal is given to the input of the pulse

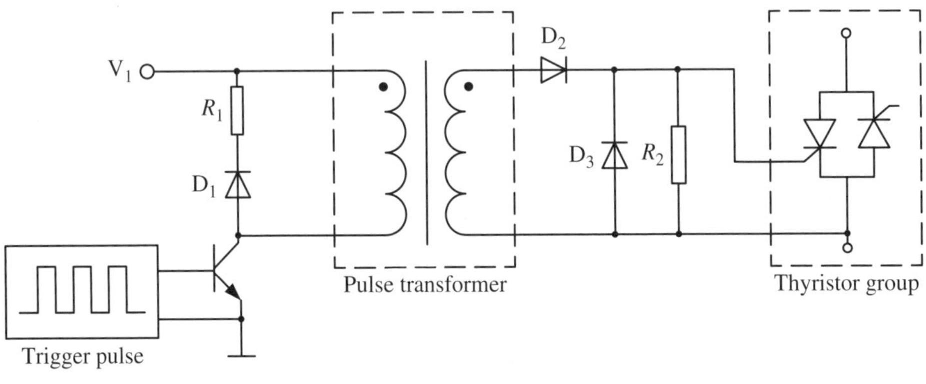


**Figure 7.16** Thyristor sync pulse trigger circuit.

transformer after amplification. The pulse signal is enhanced by the pulse transformer, forming a strong trigger signal, then output to the SCR gate, thereby triggering the thyristor reliably. Thyristor sync pulse has a total of six triggering circuits, and every two roads driving a group of thyristor.

#### 7.4.1.2 The Principle of the Thyristor Phase-Controlled Reactive Power Regulation

The method of the thyristor phase-control reactive regulation is assuming each acquisition time is $\Delta t$, namely step size. The reactive power of each phase which is detected in the first $\Delta t$ is $\Delta Q$, and compensating the reactive power $\Delta Q$ in the next $\Delta t$. Thyristor phase-control reactive power regulation is regulating the voltage on the capacitor. The capacitor voltage is the chopping voltage which is related to the firing angle that is nonsinusoidal voltage. At this point, the capacitive reactive power is

$$Q_c = \omega C U_\alpha^2 \tag{7.30}$$

where $U_\alpha$, the nonsinusoidal RMS voltage when the trigger angle is $\alpha$.

Assuming the compensated reactive power $\Delta Q$ is equal to $\Delta Q_c$ that chopper voltage generated at the capacitor within the time $\Delta t$. Namely

$$\Delta Q = \omega C U_\alpha^2 \tag{7.31}$$

Defined by voltage RMS, namely

$$U_\alpha = \sqrt{\frac{1}{\pi}\int_\alpha^\pi U_m^2 \sin^2(\omega t)\mathrm{d}(\omega t)} \tag{7.32}$$

Formula (7.32) is substituted into formula (7.31), which can be pushed into a relationship between the trigger angle $\alpha$ and the reactive power

$$\sin 2\alpha - 2\alpha = \frac{4\pi \times \Delta Q}{\omega C U_m^2} - 2\pi \tag{7.33}$$

where $\omega$, the power angular frequency;
$C$, the value of the capacitor;
$U_m$, the amplitude voltage of the chopper.

This is a transcendental equation about trigger angle $\alpha$, and it is very difficult to obtain the expression of the angle $\alpha$ directly. First, we can obtain the reactive power corresponding to the trigger angle taken $\alpha = 0°, 2.5°, 5°, \ldots, 180°$, for example, which corresponds to the reactive power $\Delta Q = \Delta Q_1, \Delta Q_2, \ldots, \Delta Q_{72}$, respectively, and then calculate the firing angle $\alpha$ according to the interpolation between the two measured points reactive power.

Reactive power adjustment step $\Delta t$ can be determined by the load change cycle. The faster the load changes, the smaller the step size $\Delta t$ is; the slower the load changes, the larger the step size $\Delta t$ is.

As seen in Figure 7.14, this dynamic compensation is lag tracking control which is an open loop control. Within a step size (e.g., $\Delta t = 100\text{ms}$) detect the reactive power of the motor, then compensate just the detected value in the beginning of the next $\Delta t$, simultaneously detected reactive power in $\Delta t$. Based on instantaneous reactive power, to calculate the firing angle $\alpha$ and the phase of the six trigger pulses, the six thyristors are triggered by the six pulses issued by the microcomputer timely.

There are three issues we should pay attention to in the control process.

1) When system starts, conduction angle $\beta$ is gradually increasing. Before powered on, the capacitor has no charge; in the first instant of the trigger pulse, despite the limited flow reactor, the thyristor will have a greater impact current. Therefore, after power on, the conduction angle $\beta$ is gradually increased, the impulse current can be effectively suppressed.
2) Determine cycle reasonably. As can be seen from the above analysis, the smaller the step size is, the better the tracking compensation. A thyristor phase control cycle cannot be less than 20 ms. During one or two cycles, the stability of instantaneous reactive power is relatively poor. Therefore, several or a dozen-frequency cycles are usually selected as a regulatory step.
3) If in a major cycle, the load fluctuation is very large, large cycle time is relatively short, such as 10 s or tens of seconds, so the reactive power changes are very complex (see Figure 7.22). In this case, we can first identify the reactive power curve, and then adjust the step size and the compensation amount according to the appropriate curve.

### 7.4.2 Dynamic Compensation-Based IGBT Control

As mentioned above, the control cycle of dynamic reactive power regulation based on thyristor phase control cannot be less than 20 ms, so the performance of the reactive power compensation is relatively poor. As we all know, the IGBT switching frequency is much higher than the thyristor, and therefore higher smoothness requirements for reactive power compensation, can use IGBT control dynamic reactive power compensation. IGBT control based on the motion compensating circuit is divided into two: one is a voltage-type bridge circuit, and the other bridge circuit is a current type. The difference between the DC sides is that it respectively uses the capacitance and inductance of these two different storage elements. For a voltage bridge circuit, a series reactor is needed to contribute to the power grid. In the current bridge circuit, a shunt capacitor is required in order to contribute to the power grid. In practical application, SVG is mostly used the as bridge circuit of the voltage type. Therefore, SVG specifically refers to the use of self-commutated voltage-type bridge circuit as a dynamic reactive power compensation device. The IGBT-based dynamic compensation circuit is shown in Figure 7.17. The voltage-type bridge circuit control system includes voltage and current detecting circuit, a phase detection circuit, detection control circuit of the microcontroller, and the PWM wave modulation circuit.

#### 7.4.2.1 Circuit Schematic Based on IGBT Dynamic Compensation

In Figure 7.17, the voltage, current, and phase detection circuit is identical to Figure 7.14; the microcontroller detects voltage, current, phase voltages, and phase currents between the retardation of the method is identical to Figure 7.14. When the SVG system works properly, the microcontroller detects the motor voltage, and current, and using formula (7.29) obtains the reactive power. The microcontroller circuit sends out the PWM drive

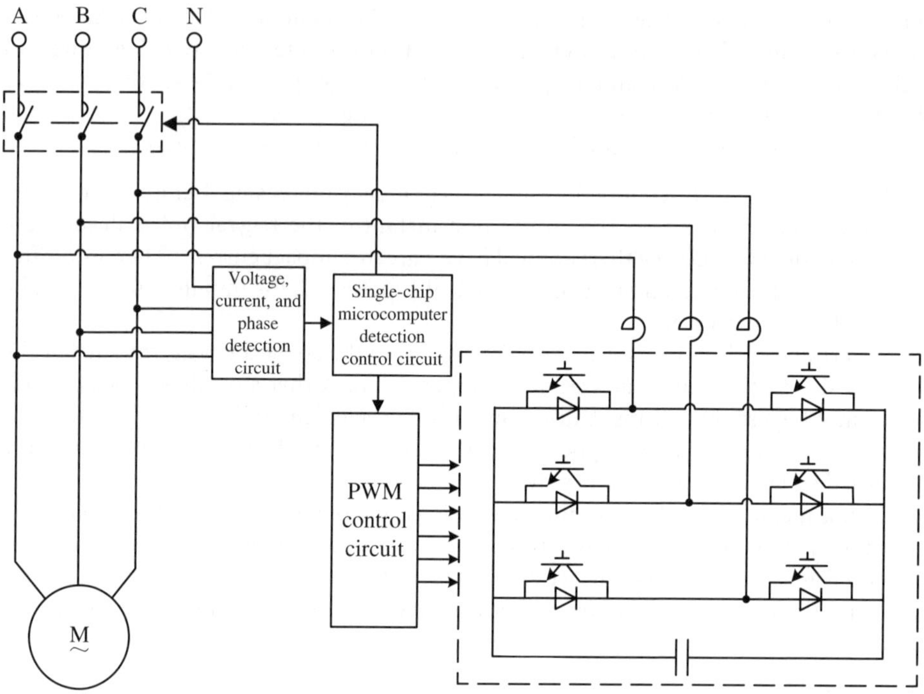


**Figure 7.17** The voltage-type bridge SVG circuit schematic block diagram.

signal to control the IGBT of the PWM circuit. The PWM circuit inverts the DC voltage into AC sinusoidal voltage with the same frequency of the power grid. The output sinusoidal voltage goes through the smoothing reactor filter and compensates the reactive power of the grid and load. The harmonics of the SVG output sinusoidal voltage can be ignored because of the filter, so only the fundamental component needs to be dealt with.

#### 7.4.2.2 Theory of Reactive Power Regulation Based on IGBT

As SVG output the same frequency AC voltage with the grid in normal work by the bridge inverter circuit, the SVG can be seen as a controlled source and AC power grid is simplified as a voltage source. The connection impedance between it and SVG can be equivalent to a reactor; moreover, the single phase can be obtained at this time, and the equivalent circuit can be shown in Figure 7.18a. The voltage on the reactor is the difference between the vector of $\dot{U}_s$ and $\dot{U}_1$, current on the reactor can be controlled by $\dot{U}_L$, changing the output voltage of SVG $\dot{U}_1$ and the phase, you can change $\dot{I}$ and the nature and size of SVG absorption reactive power.

In the circuit shown in Figure 7.18, the reactor is considered as a pure inductance that is excluding its loss and inverter losses. Therefore, it is not necessary to consider the functional capacity of the SVG which is absorbed from the grid. In this case, $\dot{U}_1$ is simply the same phase as $\dot{U}_L$; only changing the amplitude of $\dot{U}_1$, you can control the current of SVG which is ahead or behind of 90° and control the amplitude of the current. When $U_1$ is less than $U_s$, the current lag system voltage 90°, SVG absorbs the inductive reactive power from the system, as shown in Figure 7.18b. When $\dot{U}_1$ is more than $\dot{U}_s$, the current

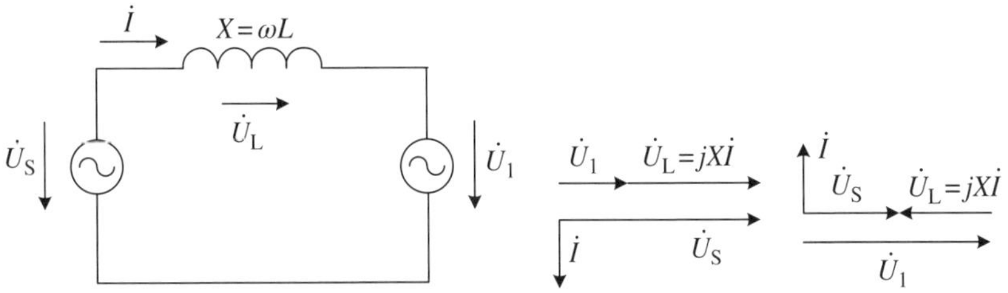


**Figure 7.18** SVG equivalent circuit and phasor graph. (a) Equivalent circuit, (b) absorb the perceptual reactive power, and (c) absorb the capacitive reactive power.

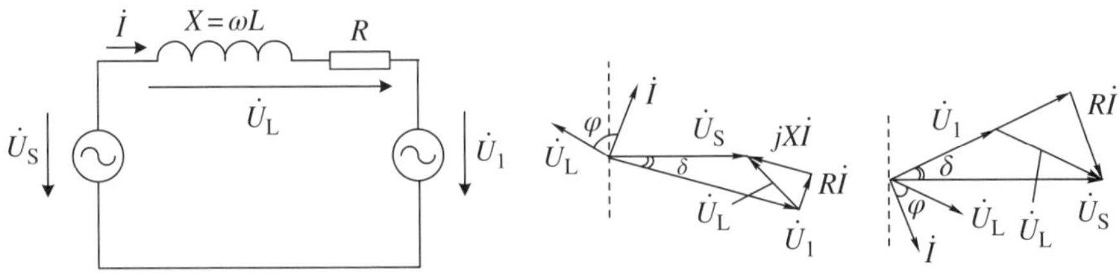


**Figure 7.19** SVG equivalent circuit and the phasor graph (the loss including). (a) Equivalent circuit, (b) current advance, and (c) current hysteresis.

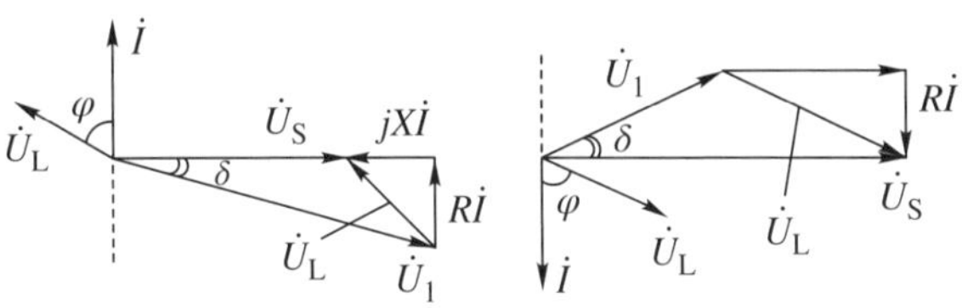


**Figure 7.20** SVG equivalent circuit and the phasor graph (the loss excluding). (a) Current advance and (b) current hysteresis.

will be ahead of system voltage 90°. SVG absorbs capacitive reactive power from the system, as shown in Figure 7.18c.

When the loss of the reactor and loss of the current transformer itself (such as the tube voltage drop, the line resistance losses, etc.) are considered, SVG total power loss can be equivalent to the resistance loss of the reactor and the SVG is the actual equivalent circuit shown in Figure 7.19a. Because of the resistor, the current lead and lag the vector diagram of Figure 7.19b, c. At this time, the phase difference between the voltage and current of the inverter is still 90°, while the phase between the grid voltage and current is no longer 90° but smaller than 90°, so the active power grids to supplement the loss of the SVG circuit, that is, speaking with respect to the grid voltage, current, a certain amount of active component. Change the angle, and change the amplitude of the phase, and magnitude of the current generated will change. Therefore, SVG will be adjusted.

In addition, there is a loss of energy program provided by the DC side in practical engineering, which is different from scheme above. In this case, the work phasor graph is different from Figure 7.19. Grid voltage differs from the current phase by 90°, while the phase between the voltage and current of the inverter reach $90° + \delta$, as shown in Figure 7.20.

In summary, compared with the traditional representation of TCR, SVG can adjust faster, have a wide operating range, and after taking multiple or PWM technology measures, it can greatly reduce the content of compensation current harmonic. More importantly, the SVG uses reactance and capacitive element smaller than SVC who uses a reactor and a capacitor; it will greatly reduce the size and cost of the device. Because SVG has such a superior performance, it is a very important development direction for the dynamic reactive power compensation device in future. However, there are also some shortcomings in SVG: on the one hand, its control method and control system are more complex than traditional SVC; on the other hand, you will use a large number of self-turn-off devices with high-capacity, and it costs much higher than the ordinary thyristor. Therefore, SVG not only has the potential overall cost advantage by the small energy storage element, but also need to be promoted by the use of devices raising the level and cost reduction.

In addition, two dynamic compensation methods above are open-loop tracking control and SVG regulation of speed, and reactive compensation effect are better. The speed of adjustment of SVC is slow and reactive compensation effect somewhat less certain. In general, in the reactive power compensation situation, the requirements are not high, SVC and SVC compensation device costs low, and of high reliability. The reactive power compensation accuracy is demanding and requires a closed-loop control. The closed-loop control is necessary to increase the set of voltage and current detection circuit and phase detection circuit, and greatly increased costs.

## 7.5 Hybrid Compensation

Dynamic reactive power testing of the motor with the change of the load cycle shows: changes in the active and reactive power of the motor are periodic, and the reactive power fluctuation part of the amplitude variation is not large. Therefore, dynamic compensation can be divided into two stages: the basic part is static compensation for the dynamic compensation of fluctuations, such compensation is called hybrid compensation; the mixed compensation of the motor costs can be greatly reduced.

### 7.5.1 Fluctuation Part of the Dynamic Compensation

For the periodic load, theory and tests show that motor reactive power is greater than the no-load reactive power, less than the rated reactive power range fluctuates. For that special load, the fluctuation frequency is relatively high. For instance, a station pumping motor in the oil field was tested. Its active power curve is shown in Figure 7.21, reactive power curve is shown in Figure 7.22, and the changes in the power factor curve is shown in Figure 7.23. The model of the three-phase induction motor Y280M-8, rated power of 55 kW, rated voltage 380 V, and rated current 113.5 A.

Pumping fluctuations in the load seen from Figure 7.21 is always a period of time within a cycle in the generator state, the maximum load power is not more than 14 kW, the maximum power generation is about 4 kW (4000 W). This part of the power generation is a feedback grid of power. Within one cycle, pumping is always a period of a power generation state, a different state as well, the generating capacity is not the same. As can be seen from Figure 7.22, the reactive power is fluctuating, fluctuations in the

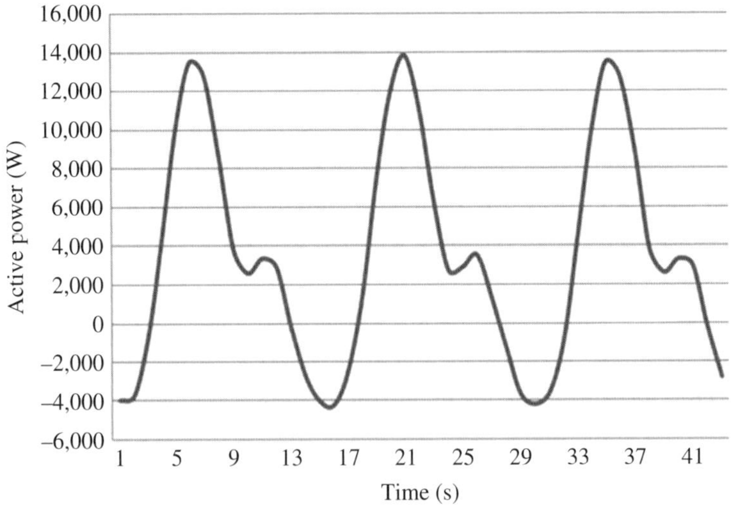


**Figure 7.21** Pumping unit motor active power curve.

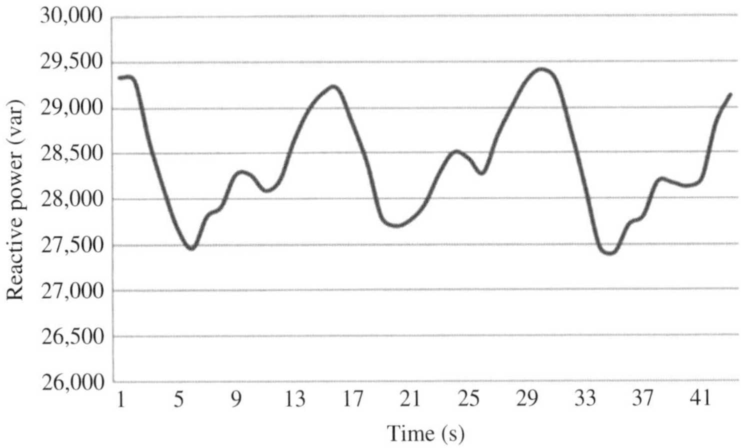


**Figure 7.22** Pumping unit motor reactive power curve.

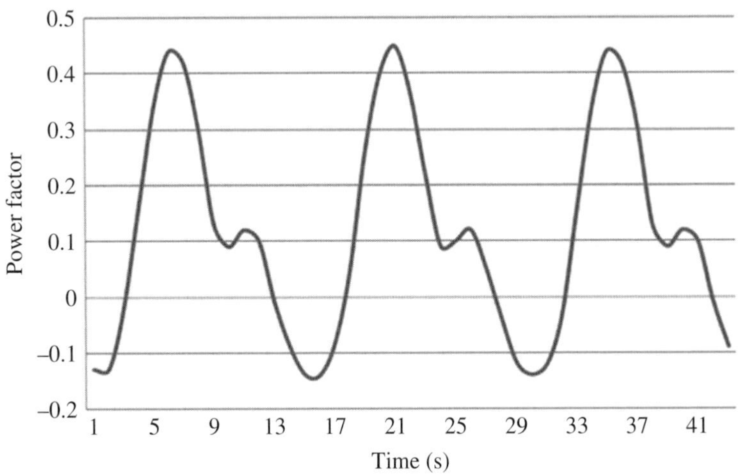


**Figure 7.23** Pumping unit motor power factor curve.

range between 27.5 and 29.5 kvar. The average is about 28.5 kvar. As can be seen from Figure 7.23, the fluctuations in the power factor are large, the fluctuations in the range are −0.15–0.45. Power generation, the power factor is negative, indicating that the current and voltage angle greater than 90°. Figures 7.21, 7.22, and 7.23 show the three curves and the more comprehensive reflection of the pumping unit load characteristics.

For the complexity load of the pumping unit, if you require a higher power factor after compensation, you may use method of hybrid compensation. The so-called hybrid compensation is using static way for the main part of the reactive power compensation, using a dynamic way for the part fluctuation of the reactive power compensation. In the curve shown in Figure 7.22, the reactive power, for example, in the main part with a 27 kvar capacitor compensation curve is used to compensate for the fluctuation part of the reactive power in Figure 7.24, and then through a thyristor controlling capacitor (2.5 kvar).

A hybrid compensation circuit is shown in Figure 7.25, the $C_J$ in the left is a static compensation capacitor which is used to compensate the lower portion of the straight and the $C_d$ in the right is the dynamic compensation capacitor which is used to compensate for the shaded portion above the line. Because the static compensation compensates most of the reactive power, the dynamic compensation capacitor $C_d$ is small. The thyristor current is also very small and the cost is greatly reduced. Take the test of motor above, for example, if the full advantage of the dynamic compensation is taken, line current of the transistor compensating devices is

$$I_c = \frac{Q_c}{\sqrt{3}U_N}$$

When $Q_c = 29.5\text{kvar}$,

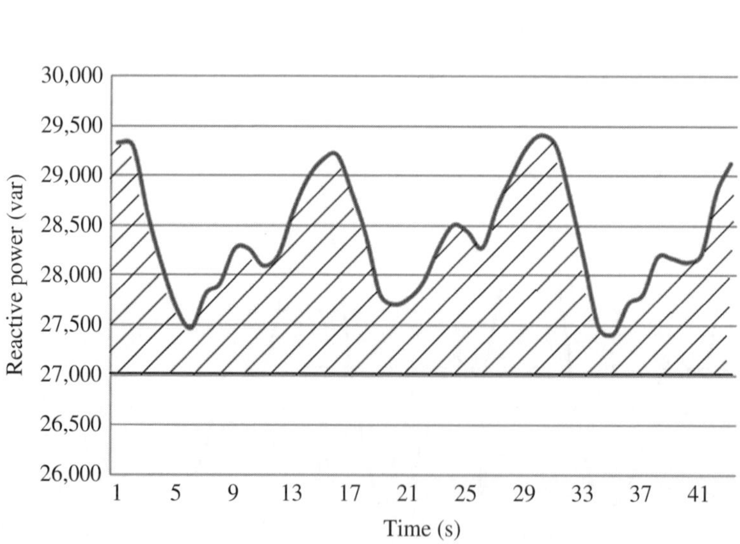


**Figure 7.24** Fluctuation part of the dynamic compensation schematic diagram.

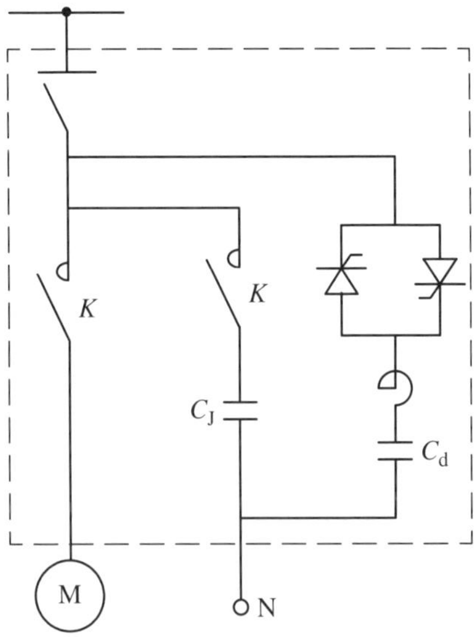


**Figure 7.25** Fluctuation part of the dynamic compensation circuit schematic.

$$I_c = \frac{Q_c}{\sqrt{3}U_N} = \frac{29.5\times 10^3}{\sqrt{3}\times 380} = 44.8\,\text{A}$$

When $Q_c = 2.5\text{kvar}$,

$$I_c = \frac{Q_c}{\sqrt{3}U_N} = \frac{2.5\times 10^3}{\sqrt{3}\times 380} = 3.8\,\text{A}$$

As can be seen from the above calculation result, the circuit shown in Figure 7.25, the thyristor current is reduced. By selecting the thyristor with small current, the cost will be reduced a lot. A large power motor need to be compensated dynamically, the hybrid compensation can greatly reduce costs and improve reliability. This dynamic compensation can not only greatly reduce the cost, but once part of dynamic compensation does not work, static compensation part can meet the basic requirements of the motor reactive power compensation.

### 7.5.2 Over Make Up Part of the Dynamic Compensation

In the above example, you can also take static compensation first and then take dynamic compensation. That is, the first method adopts a 29.5 kvar capacitor to fill the peak of the reactive power, then control inductive current through the thyristor, that is, control the inductance dynamically complementing reactive power absorption (Figure 7.26; shade portion). This kind of dynamic and static mixing compensation circuit is shown in Figure 7.27. Compared with mixed compensation as shown in Figure 7.25, the cost of this hybrid compensation is increasing, but reliability is high and the compensation effect is good.

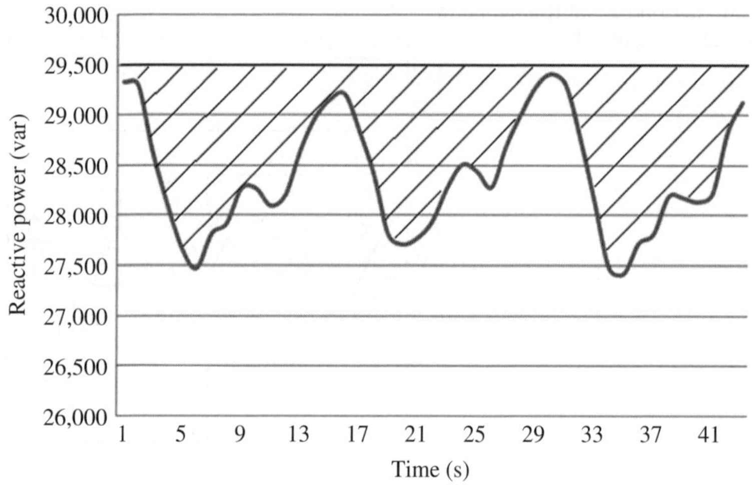


**Figure 7.26** For the dynamic compensation part of the diagram.

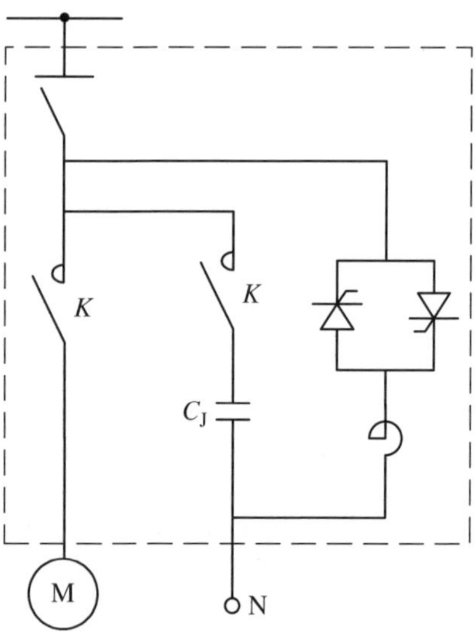


**Figure 7.27** Circuit of the dynamic compensation.

## 7.6 The Discussion Topic of the Chapter

1) Why do you use three-phase induction motor reactive power compensation on the spot? What are the benefits?
2) How many species of reactive power compensation are there?
3) How can you calculate rated load and no-load: two cases of reactive power?
4) From reactive power theory analysis, calculate how many components are involved in the motor reactive power?
5) What kind of role does reactive power compensate on the spot?

6) How to calculate reactive power compensation capacitor in place?
7) From the viewpoint of the current phasor diagram, illustrate the energy-saving principle of reactive power compensation.
8) What circumstances require dynamic reactive power compensation?
9) To achieve dynamic reactive compensation, what are the advantages and disadvantages of the thyristor?

# Further Reading

Sun Chengbao, Jin Zhe. Modern Energy-Saving Technology and Energy-Saving Engineering. Beijing: China Water Power Press, 2005.
Fang Daqian. Pump, Fan and Crane Speed Check Manual. Beijing: China Water Power Press, 2009.
Yao Fulai, Zhang Yanfang, et al. Electrical Energy Saving Control Method and Practice. Beijing: China Electric Power Press, 2009.
Dai Guangping, Han Bing. The Principle of Motor Reversibility and Motor Energy Saving Energy. Beijing: China Petrochemical Press, 2008.
Huang Guozhi, Fu Fengli. Y2 Series Three-Phase Asynchronous Motor Technology Manual. Beijing:China Machine Press, 2005.
Zhao Jiali. A Question-and-Answer Session of Motor Fuel Economy. Beijing: Chemical Industry Press, 2008.
Liu Lijun. Energy Saving Technology and Energy-Saving Engineering. Beijing: China Electric Power Press, 1999.
Liu Lijun. Energy Saving Technology and Its Engineering Applications. Beijing: China Electric Power Press, 2011.
Yu Longhai. Motor Energy Efficiency and Energy-Saving Technology. Beijing: China Machine Press, 2008.
Motor Project Team. Energy Efficiency and Market Potential of Electric Motor System in China. Beijing: China Machine Press, 2001.
Zhao Sheng, Zhao Kai. Practical Guide for Energy Saving of Electric Motors. Beijing: China Machine Press, 2009.
Feng Songjing. Electromechanical Equipment Technology and Energy Saving. Xi'an: Northwestern University Press, 2004.
The Editorial Committee of the Technical Manual of Practical and Electronic Energy Saving Technology. Practical Mechanical and Electrical Energy Saving Technology Manual. Beijing: China Machine Press, 1997.
Zhao Xizhang. Energy Saving Technology and Methods. Beijing: China Machine Press, 2004.
Wang Zhanyuan, Xiao Shuxia, et al. Practical Windings of Common Motors. Beijing: China Machine Press, 2006.
Yao Zhisong, Wu Jun. An Example of Motor Energy Saving Method and PLC Converter. Beijing: China Electric Power Press, 2010.

*Energy-saving Principles and Technologies for Induction Motors*, First Edition. Wenzhong Ma and Lianping Bai.

# Index



*Energy-saving Principles and Technologies for Induction Motors*, First Edition. Wenzhong Ma and Lianping Bai.

### w



### y